M50

Water Resources Planning

Third Edition

American Water Works Association

Manual of Water Supply Practices—M50, Third Edition

Water Resources Planning

Disclaimer

The authors, contributors, editors, and publisher do not assume responsibility for the validity of the document or any consequences of its use. In no event will AWWA be liable for direct, indirect, special, incidental, or consequential damages arising out of the use of information presented in this book. In particular, AWWA will not be responsible for any costs, including, but not limited to, those incurred as a result of lost revenue. In no event shall AWWA's liability exceed the amount paid for the purchase of this book.

Project Manager/Managing Editor: Melissa Valentine
Production: Glacier Publishing Services, Inc.
Manual Specialist: Sue Bach
Cover Design: Melanie Yamamoto

If you find errors in this manual, please email books@awwa.org. Possible errata will be posted at www.awwa.org/resources-tools/resource.development.groups/manuals-program.aspx.

Library of Congress Cataloging-in-Publication Data

Names: Kammereck, Lara, author. | American Water Works Association, issuing body.
Title: Water resources planning / by Lara Kammereck.
Other titles: AWWA manual ; M50.
Description: Third edition. | Denver, CO : American Water Works Association, [2016] | Series: [AWWA manual] ; M50
Identifiers: LCCN 2016019686 | ISBN 9781625761798
Subjects: LCSH: Water-supply--Planning. | Water consumption--Forecasting. | Water resources development.
Classification: LCC TD345 .K268 2016 | DDC 363.6/10684--dc23 LC record available at https://lccn.loc.gov/2016019686

ISBN-13 978-1-62576-179-8
eISBN-13 978-1-61300-390-9

Printed in the United States of America
American Water Works Association
6666 West Quincy Avenue
Denver, CO 80235-3098
awwa.org

Contents

Figures

Tables

This page intentionally blank.

Preface

This publication is the third edition of the American Water Works Association (AWWA) Manual M50, *Water Resources Planning*, originally published in 2001, with a second edition released in 2007. The manual presents information on how to develop a plan for new water supplies to accommodate projected future water demands and provides an effective framework for water resources planning for the 21st century.

Water resources planning for potable water supplies is a very broad topic. Issues range from estimating future water demand to evaluating possible new sources of water, protecting water sources, and addressing expanding environmental regulations. One method for preparing a water resources plan is integrated resource planning (IRP). Developed in the 1990s, IRP is a way to bring together myriad issues, interests, and stakeholders through a planning process that can result in a reason-based, cost-effective, and environmentally sound plan the public can support.

This third edition significantly enhances the basis of water resources planning provided in prior editions. Additions and improvements include the following:

- Organization of the manual around the planning process
- Inclusion of stakeholder involvement at various steps throughout the process
- Inclusion of reclaimed water and other source water alternatives in the portfolio of solutions, recognizing the Total Water Solutions planning perspective
- References to AWWA standards and manuals of water supply practices for more details on specific topics
- Expanded emphasis on the evaluation of alternatives
- A new, award-winning case study that illustrates the concepts presented in the manual
- Inclusion of the AWWA policy statement on water resources planning

This manual was prepared by AWWA's Water Resources Planning and Management Committee and associated volunteer authors to help water resources planners meet the challenge of accommodating a growing demand for water, while complying with myriad regulations that govern the development and use of new water supplies.

No two water resources plans are alike, and planning is an ever-changing process. The issues discussed in this manual should allow the planner to develop and implement a comprehensive work plan that responds to technical and institutional questions that must be addressed before deciding how to develop new water supplies.

The AWWA Water Resources Planning and Management Committee welcomes your input and comments on the content and usefulness of this manual.

This page intentionally blank.

Acknowledgments

The AWWA Technical and Educational Council, the Water Resource Sustainability Division, and the Water Resources Planning and Management Committee gratefully acknowledge the contributions made by those volunteers who drafted, edited, and provided the significant and critical commentary essential to updating AWWA Manual M50. The Technical Review Board members dedicated countless hours in the final stages of preparation of this edition to ensure the overall technical quality, consistency, and accuracy of the manual.

Technical Review Board Members

Lara Kammereck, Chair, Carollo Engineers, Seattle, Wash.
Frederick Bloetscher, Florida Atlantic University, Hollywood, Fla.
William Y. Davis, CDM, Carbondale, Ill.
Thomas E. Dumm, O'Brien & Gere, Bowie, Md.
Chi Ho Sham, Eastern Research Group, Inc., Waltham, Mass.
Brian Skeens, CH2M, Atlanta, Ga.

Contributors to the 3rd Edition

Lara Kammereck, Chair, Carollo Engineers, Seattle, Wash.
John Andrew, California Department of Water Resources, Sacramento, Calif.
Frederick Bloetscher, Florida Atlantic University, Hollywood, Fla.
Samia R. Broadaway, Baker Botts L.L.P., Austin, Texas
David B. Campbell, Schnabel Engineering, West Chester, Pa.
Brad Castleberry, Lloyd-Gosselink Attorneys, Austin, Texas
Danusha S. Chandy, Arcadis, Arlington, Va.
William Y. Davis, CDM, Carbondale, Ill.
Thomas E. Dumm, O'Brien & Gere, Bowie, Md.
Steven A. Friedman, HDR, Riverside, Calif.
Danielle Gallet, Center for Neighborhood Technology, Chicago, Ill.
Lindsey Geiger, American Water Works Association, Denver, Colo.
Alex Gerling, American Water Works Association, Denver, Colo.
Stephanie Shamblin Gray, HDR Engineering, Inc., Irvine, Calif.
Jeffrey W. Henson, Black & Veatch, Kansas City, Mo.
Pamela P. Kenel, Black & Veatch, Gaithersburg, Md.
Michelle Klase Robinson, Dialogue Public Relations, Safety Harbor, Fla.
Les K. Lampe, Black & Veatch, Kansas City, Mo.
Sharon C. Long, University of Wisconsin, Madison, Wis.
Lisa A. Maddaus, Maddaus Water Management, Folsom, Calif.
William O. Maddaus, Maddaus Water Management, Alamo, Calif.
James W. Miller, Everett Public Works Department, Everett, Wash.
Robert Morgan, Beaver Water District, Lowell, Ark.

Kelly Mott Lacroix, Water Resources Research Center, Tucson, Ariz.
Lela Perkins, Black & Veatch, Centennial, Colo.
Robert S. Raucher, Stratus Consulting, Inc., Boulder, Colo.
Klinton A. Reedy, Black & Veatch, Centennial, Colo.
John Rehring, Carollo Engineers, Broomfield, Colo.
Steven P. Roy, Geosyntec Consultants, Portsmouth, N.H.
Chi Ho Sham, Eastern Research Group, Inc., Waltham, Mass.
Brian M. Skeens, CH2M, Atlanta, Ga.
Lena Tam, East Bay Municipal Utility District, Alameda, Calif.
Jacob G. VandenBosch, Wilmington, N.C.
Janine B. Witko, Arup, New York City, N.Y.
Benjamin A. Wright, Hazen and Sawyer, Baltimore, Md.

Chapter 1

Introduction to Water Resources Planning

A number of years ago, archeologist Bryan Fagan addressed attendees at the American Water Works Association's (AWWA's) Sustainable Water Sources Conference. Dr. Fagan has traveled all over the world to study ancient ruins. He likened his career to "50 years of studying drainage ditches," and explained that through studying drainage ditches he could provide the audience with an understanding of the rise and fall of civilizations through history. His book *Elixir* outlines a number of these civilizations: Egypt, Babylonia, Southeast Asia, and even the American West (Fagan 2012). He found that a civilization could develop to the extent that infrastructure could be constructed to allow water to flow to where it was needed for agricultural purposes, whether that was Alexandria or Ur. If the water stopped flowing, there was trouble.

As a result, later civilizations expanded and developed as technology allowed water to flow farther to protect themselves from drought or other outages. Rome demonstrated that water could be moved with more than ditches, which would have been a severe limitation. The Romans constructed extensive tunnels and aqueducts to supply Rome with water from mountains to the east and north, some of which are still in operation. Some people have estimated that we know perhaps only about 20 percent of the Roman tunnel system in that we keep discovering more of it each year—the location of many of the tunnels in the system were lost in the Dark Ages after the fall of Rome.

Dr. Fagan demonstrated that it was access to water that allowed human civilizations to develop and evolve. It is why a number of engineering organizations like Water For People and Engineers Without Borders focus their efforts on providing access to clean water to people in Third World countries. It is critical for people to remain healthy enough to be productive in the modern world. Other infrastructure (roads, major buildings, etc.) also depends in part on people having access to adequate clean water supplies, allowing our communities and economies to thrive.

Water availability, however, can be a limiting factor. Although 70 percent of the earth's surface is covered by water (mostly in oceans), only about 3 percent of all water on earth is freshwater (USGS 2015). Because much of this freshwater is locked up in ice caps and glaciers, only about 1 percent of all freshwater is reasonably accessible for use.

Far less than 1 percent of the world's total supply of water is considered easily accessible for human use.

While supplies of freshwater are finite, its uses are not. Now estimated at more than 7 billion, the world's population is increasing by some 70 to 80 million people per year. By 2025, the number will exceed 8 billion. Hundreds of millions of people around the world already face critical water shortages. Some areas of the United States also face shortages, and other areas are at risk of future shortages. That is where water resource management and the concept of sustainability come into play. The concept of sustainability reflects water consumption at a rate that can be maintained in perpetuity, while maintaining water quality.

But the concept is not that simple, and buy-in to the concept of "sustainable water" depends on the profession or perspective of the person defining it (Bloetscher 2010). A sustainable world is not a rigid one where population or productivity is constant. Instead, the concept of sustainability must adapt to constant change. The concept of sustainability requires rules, laws, and social constraints that are recognized and adhered to by all (Meadows, Randers, and Meadows 2004). From a hydrologic perspective, the term "sustainable yield" is the amount of water that can be withdrawn from a source at rates that are less than their recharge potential and that do not deteriorate the source or basin. While many areas attempt to develop and use water sustainably, not all do. Even within regionally sustainable situations, unsustainable use can occur locally.

A key component of water resource management is planning the utilization of water supplies, which requires some means to determine how the hydrologic cycle provides water to the area (e.g., recharge basin), in what quantities, and with what reliability. Ultimately, reliability is a risk issue—precipitation can be consistent or can have significant fluctuations that disrupt ongoing basin development (Molak 1997). In any given basin, there are typically a variety of uses competing for water resources, and each basin has unique characteristics (Bloetscher and Muniz 2008). Major uses of water can include

- agriculture,
- ecosystems,
- recreation,
- urban and domestic use, and
- industrial and power generation use.

All these users depend on—and in many cases can compete for—reliable water supplies. From a practical perspective, sustainable development of a region depends on the appropriate management of water resources. Managing water resources requires decision-making that addresses the environmental, economic, and social concerns of the community, which may extend far beyond the community borders. The result is that researchers can define the concept of sustainability, but practitioners emphasize feasibility and limitation to sustainability of the ecosystem (Starkl and Brunner 2004).

Water quantity and quality issues have significant fiscal effect on the potential users in the basin, while there are unrealized costs and benefits that are often ignored in the current water management framework (Bloetscher and Muniz 2008). There are two categories of water supplies: surface systems and underground systems, which are described in the following sections.

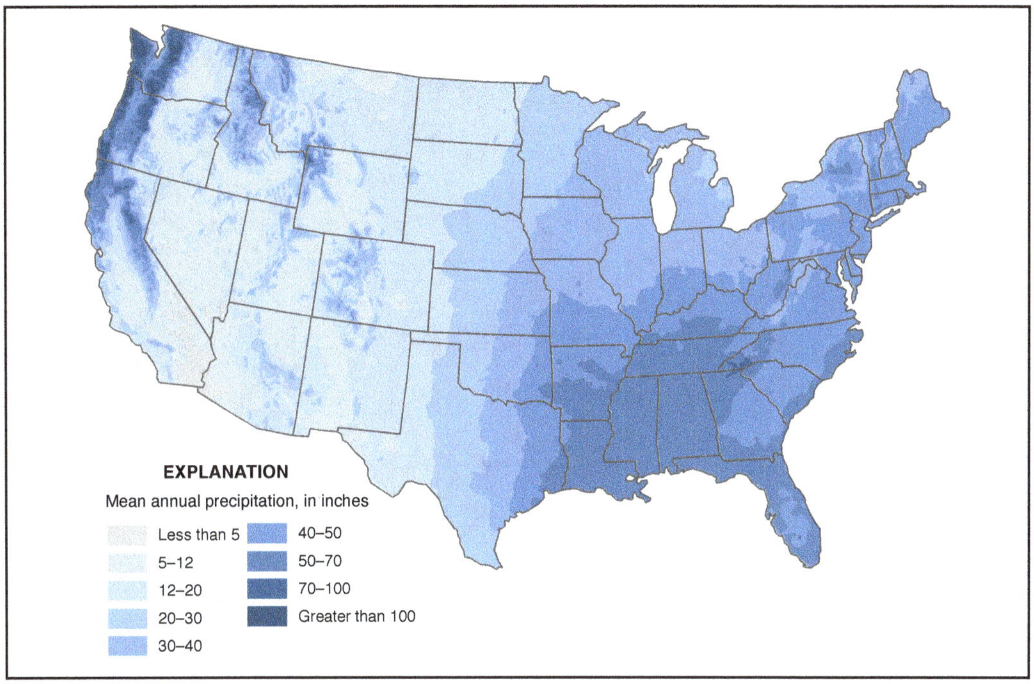

Source: *Reilly et al. 2009*

Figure 1-1 Average annual precipitation (inches) in the United States (1961–1990)

SURFACE WATER

Surface water is precipitation that runs off the land, reaching streams, rivers, or lakes, or groundwater that discharges into these water bodies. Surface water bodies that mix with saltwater bodies along the coast are called estuaries (e.g., where a river meets the ocean in a delta). Surface water flow is controlled by topography because water on the surface flows downward by gravity, which means it generally flows toward the oceans. Figure 1-1 outlines precipitation throughout the continental United States. Annual rainfall is generally much higher in the eastern United States than in the west, except in the Pacific Northwest region.

Figure 1-2 shows the difference between precipitation and potential evapotranspiration (PET). The tan areas illustrate where the PET rate is higher than precipitation, meaning net water availability (precipitation minus PET) for crops and other purposes is negative. In these areas, high levels of crop irrigation may not be sustainable, depending on local conditions and sources of supply. In many of these areas, streamflows are variable and limited, so surface water storage and groundwater are often used to ensure adequate water supplies for crops and people.

Surface waters are recharged by precipitation and sometimes by groundwater. When there is limited rain, surface waters can diminish significantly and leave users at risk. The construction of reservoirs can reduce the potential for this risk, but prolonged droughts may have severe effects on supplies in surface storage (e.g., shortages experienced in and around Atlanta in 2008). Surface water quality can also be affected by adjacent or upstream land-use activities. As a result, source water protection measures are often employed to avoid detrimental surface water quality effects from point and nonpoint sources.

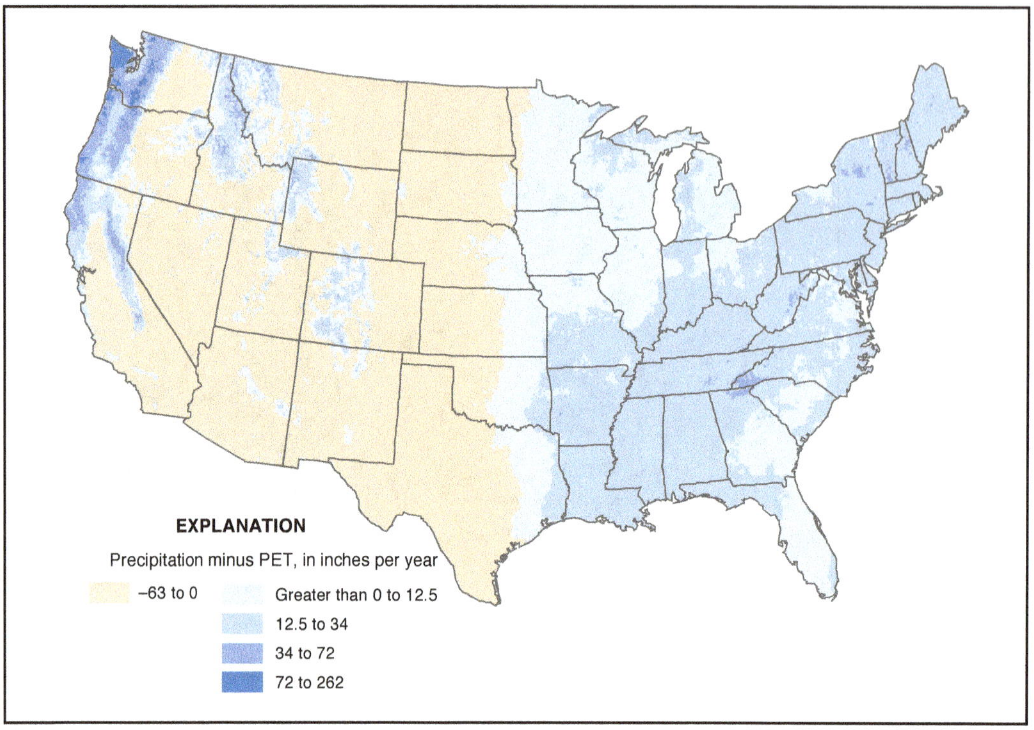

Source: *Reilly et al. 2009*

Figure 1-2 Difference between average annual precipitation and PET rates

GROUNDWATER

Groundwater is stored and flows beneath the land surface. Groundwater may be recharged by infiltration of rain and melting snow, surface waters (including streams, rivers, and lakes), and other groundwater flows. Formations that transmit groundwater are termed "aquifers." Deeper aquifers are generally recharged at slower rates relative to shallow aquifers, which are more likely to be in direct hydrologic "communication" with surface water supplies. As a result, withdrawals from deep aquifers may be more likely to be unsustainable (i.e., when withdrawals exceed recharge rates), depending on local hydrogeology, recharge rates, and patterns of use.

Ideally, planning and design of well fields should assume the wells could be pumped continuously at a rate that allows the wells to fully recharge locally. The more common practice is to pump wells that have a significant drawdown seasonally or for only a few hours each day, allowing time for the aquifer to recover. Even this is difficult, however, because the long-term capacity of an aquifer is usually determined by short-term tests and study of the geology of the area. Ultimately, using groundwater unsustainably reduces the water in aquifer storage, lowers the water table, and requires drilling and pumping from lower levels, which in turn increases capital and operating costs.

In many areas, deeper groundwater tends to have worse water quality (e.g., high total dissolved solids) as a result of having been in contact with the geologic formation longer and dissolving minerals into the water. Therefore, while some deep aquifers may be prolific, the quality of water obtained from a well may not be desirable or even usable for drinking water without substantial treatment. In addition, many deep aquifers are confined and local recharge is therefore limited. The result is the potential for aquifer drawdown that is accompanied by aquifer mining—and in extreme situations, land subsidence.

Figure 1-3 shows that the rate of groundwater recharge is very limited in much of the United States. Figure 1-4 illustrates where groundwater levels declined over the

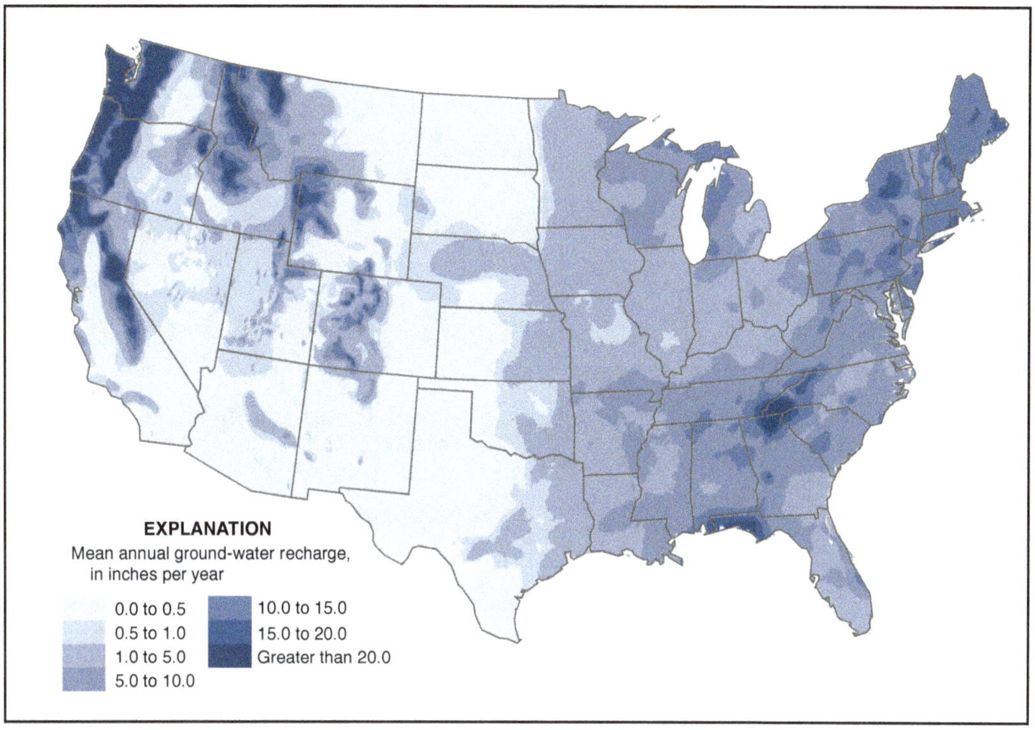

Source: *Reilly et al. 2009*

Figure 1-3 Mean annual groundwater recharge

last 40 years throughout the United States. Eastern coastal areas, the Lower Mississippi Basin states, the Great Plains states, Texas, and the West are particularly affected. Red regions indicate areas in excess of 500 mi^2 (1,295 km^2) that have water-level decline in excess of 40 ft (12 m) in at least one confined aquifer since predevelopment, or in excess of 25 ft (7.6 m) of decline in unconfined aquifers since predevelopment. Blue dots are wells in the US Geological Survey National Water Information System database where the measured water-level difference over time is equal to or greater than 40 ft (12 m). The confined aquifers recharge at very low rates, meaning that in most cases, these aquifers are being mined (overdrafted).

THE WATER RESOURCES PLANNING PROCESS

The issues discussed in this manual should allow the planner to develop and implement a comprehensive work plan that responds to technical and institutional questions that must be addressed before deciding how to develop new water supplies. However, planners are expected to use the parts of the manual that apply to their situations. Regulations and planning requirements discussed in this manual are for the United States. Specific regulations and reporting requirements of other countries should be adhered to, and the planning concepts presented in this manual adjusted accordingly.

Each chapter deals with different issues. Chapter 2 starts with the impetus for preparing a water resources plan, which can involve a variety of issues including the following:

- Changes in demand
- Changes in supply
- Raw and treated water quality
- Aging water supply infrastructure

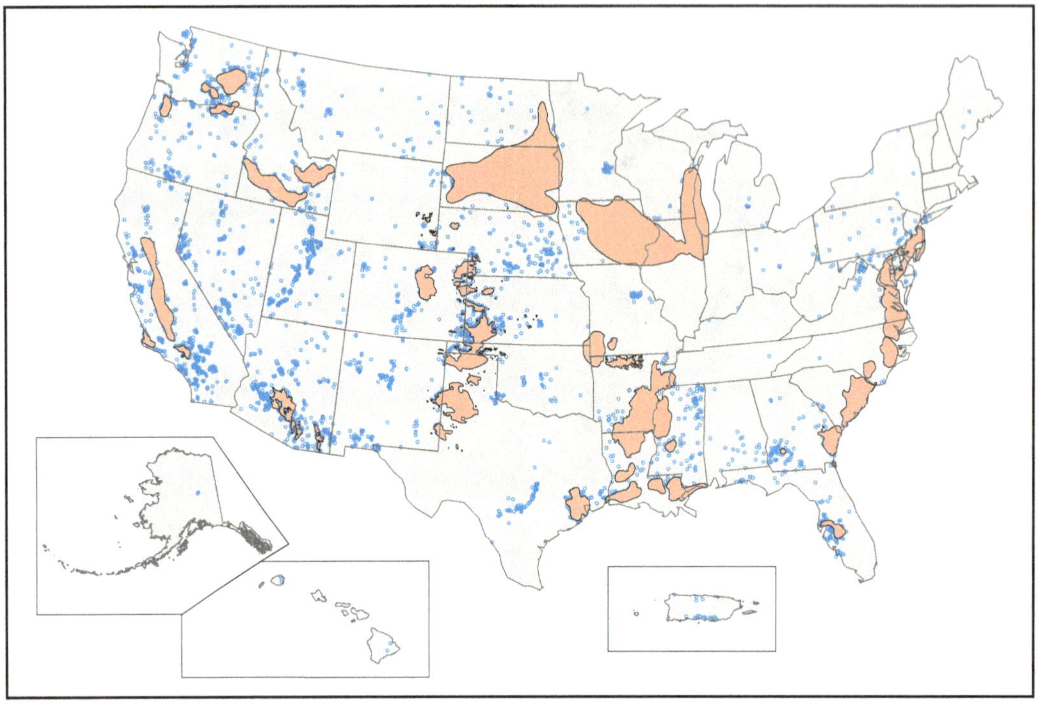

Source: *Reilly et al. 2009*

Figure 1-4 Groundwater level declines

- Implementation of conservation and reuse programs
- Instream flow and water quality requirements
- Operational changes
- Capital improvement planning

A preliminary assessment is often needed to identify issues, validate the need for the plan or a new supply, and prepare a work plan. Often, a white paper or scoping paper is needed to define the issues surrounding planning for a new source. Nearly every plan is a significant undertaking and requires approval from appointed or elected officials. The process should be thought through in advance, so decision makers have timely information to guide the process along the way and the information for selecting the best plan.

Water resources planning is ultimately a local exercise that has regional effects. The decisions associated with the planning process are critically important to the community and the community's unique local conditions because the characteristics of supply and demand are unique to each locale. The goals, values, and priorities of the end users of the water can only be coordinated through local planning processes; but because local planning is likely to have effects beyond the borders of the utility, it is important to involve other stakeholders that may affect or be affected by the plan. Active engagement by local planners in regional, watershed, and statewide planning processes is key to the success of those efforts in terms of coordinating water supply planning and heading off conflicts between competing uses of shared resources. As a result, there should be a strong and beneficial linkage between local, regional, and state planning efforts.

Chapter 2 provides an understanding of public participation techniques and the important role it can play in water resources planning. It covers the process of setting goals and developing work plans to involve the public in the water resources planning process. It also describes the key elements of plan implementation that include chartering the project team, and monitoring and managing change in the process.

Chapter 3 discusses the role of water rights in water resource management. Two distinct systems of water rights are defined and described: riparian rights, used primarily in the eastern United States; and prior appropriation doctrine, predominantly used in the western United States. Water rights systems for groundwater include absolute ownership, reasonable use, correlative rights, and appropriation permit systems. Federal and Tribal rights also are described, including federal reserved rights, water rights for water not reserved, rights of navigable water, and federal contract rights. A discussion of the water quality effects of all of these water rights doctrines is also included because this is an emerging area of interest.

Chapter 4 identifies alternatives for developing, expanding, or maximizing a water resource portfolio, including

- surface water, including source identification, screening, and site selection for direct river withdrawals, onstream reservoirs, and pumped-storage reservoirs;
- groundwater, including confined and unconfined aquifers; and
- conjunctive use of surface water and groundwater.

Chapter 4 also describes other water resources alternatives that might be available to a given community, including

- reclaimed water,
- desalination,
- demand management,
- water marketing and transfers, and
- nontraditional sources.

Chapter 4 does not necessarily present a comprehensive list of options, nor does it require evaluation of all options. Local conditions will govern the options available to the water utilities. The key is to identify feasible new sources that, individually or combined, make up the identified supply deficit for an estimated future time period.

Chapter 5 begins by summarizing common water demand forecasting methods ranging from simple per capita models to more complex statistical forecasting methods. It then describes types of data normally used for forecasting, including water billing data, weather data, demographic and economic data, and other explanatory variables, and provides examples of forecasting with per capita coefficients, land-use water factors, and regression analysis of disaggregated demand data. Uncertainty analyses and probability methods conclude the chapter.

Chapter 6 explains water quality interactions and the hydrologic cycle, including physical, chemical, and biological components. This chapter defines various measures of water quality and describes the relationships between water uses and water quality. In addition, effects of natural influences, point sources, and nonpoint sources on water quality are defined, as are the major factors affecting quality of various types of surface water. Similarly, the major factors affecting groundwater quality include common contaminants and transport considerations.

Chapter 7 addresses source water protection. Source water protection has long been promoted and used by water systems, local governments, and states as a part of a multiple-barrier approach to ensure that drinking water for public water systems is high quality and meets drinking water standards for public health protection. Source protection essentially makes treatment easier by minimizing the contaminants that must be removed. Chapter 7 includes sections on

- source water protection programs,
- groundwater protection, and
- surface water source protection.

The above protections include the following elements:

- Identification of potential contaminant sources
- Assessing vulnerability and determining susceptibility
- Making source water assessments available to the public
- Development of a source water protection plan and program

AWWA's Utility Management Standards program (also known as the G Series) is designed to cover the principal activities of a typical water and/or wastewater system. It provides a means to assess service quality and management efficiency based on recognized standards for best available practices. American National Standards Institute (ANSI)/AWWA Standard G300-14 (AWWA 2014) and the associated operational guidebook (Sham et al. 2010) are valuable resources that complement the content of chapter 7.

Chapters 8 and 9 are intrinsically linked. Chapter 8 describes water supply assessments and is designed to help the reader identify the fatally flawed options and the most likely portfolio of solutions. The solutions are compared to a baseline of performance, while risks and benefits are assessed. Infrastructure, water quality and treatment of supplies, water source reliability, environmental impacts, implementation considerations, and costs should be taken into account. The evaluation criteria should be defined in conjunction with the goals of the utility.

Chapter 9 identifies the means for assessing the evaluation criteria and utility goals through a rigorous evaluation process, reflecting the nature of the new alternative sources identified. All relevant issues regarding alternative analysis should be addressed, including

- water quality of new sources,
- protection of current and new sources,
- regulations,
- water rights and policies,
- yield of new sources and effects on other sources (assessed through modeling)
- environmental impact assessments,
- economic feasibility and financing considerations, and
- long-term viability (design life).

These evaluations are intended to identify a preferred plan that balances conflicting factors. Criteria will generally include

- cost-effectiveness,
- financial feasibility,
- public acceptability, and
- environmental issues.

Chapter 10 ties all prior chapters together by outlining the planning process and planning tools. Examples are provided to illustrate the planning process and strategies. What is not discussed in detail in chapter 10, because of constantly changing rules, are regulatory requirements. Water resource projects are subject to varying degrees of regulatory approval. Developing new sources or expanding withdrawals may require permits. Examples of such permits are a state water allocation permit or a federal Section 404 permit, often requiring a long and involved approval process.

Planners should address permit requirements in the water resources plan through consultation with state and federal regulatory officials so that likely permits are identified, and the considerations required or implied are included in the alternatives analyses. However, detailed aspects of the permit approval, such as detailed environmental impact analyses, will occur after the plan is approved. Each state or locale has its own framework for assessing the environmental impacts of water resource projects. The state's water allocation agency can identify the lead permitting agency and help identify the types of environmental work that will likely need to be performed. Detailed environmental assessments or impact statements are not usually undertaken until after the preferred alternative source is identified. However, the planning process leading to that selection should include preliminary consideration of all issues likely to require detailed assessment.

Preliminary environmental screening needs to be integrated with the alternative assessment process. This will help justify the alternative selection and reduce the time required to complete environmental assessments or an environmental impact analysis for the recommended project and needed permits due to "false starts" and missing the obvious.

SELECTING THE PROJECT TEAM

One of the most important issues a water resources planner faces is who should be involved in the planning process. Those involved are called "stakeholders" because they have a vested interest in how the process turns out and which alternative is selected. No longer are plans prepared in a vacuum, released for approval after all the technical work is completed, and the alternatives evaluated. Planners have found that this closed-door approach necessitates starting over when opposition to the proposed project surfaces, possibly more than once. An open approach requires a slower, more deliberate process. Although involving all stakeholders may slow down the technical work, the final decision and project approval are invariably reached sooner and more amicably. This critical element is discussed in more detail in chapter 2.

REFERENCES

AWWA (American Water Works Association). 2014. ANSI/AWWA Standard G300. *Source Water Protection*. Denver, Colo.: AWWA.

Bloetscher, F. 2010. The impact of unsustainable ground water. In *Florida Water Resource Conference Proceedings, May 16–19, 2010, Orlando, Fla.* Kissimmee, Fla.: FWRC.

Bloetscher, F., and A. Muniz. 2008. Water supply in South Florida—The new limitations. In *Sustainable Water Sources (Reno) Conference Proceedings*. Denver, Colo.: AWWA.

Fagan, B. 2012. *Elixir*. New York: Bloomsbury Press.

Meadows, D., J. Randers, and D. Meadows. 2004. *Limits to Growth: The 30-Year Update*. White River Junction, Vt.: Chelsea Green Publishing.

Molak, V. 1997. *Fundamentals of Risk Analysis and Management*. Boca Raton, Fla.: Lewis Publishers.

Reilly, T.E., K.F. Dennehy, W.M. Alley, and W.L. Cunningham. 2009. *Ground-Water Availability in the United States*. USGS Circular 1323. Reston, Va.: US Geological Survey.

Sham, C.H, R.W. Gullick, S.C. Long, and P. Kenel. 2010. *Operational Guide to AWWA Standard G300, Source Water Protection*. Denver, Colo.: AWWA.

Starkl, M., and N. Brunner. 2004. Feasibility versus sustainability in urban water management. *Journal of Environmental Management*, 71:245–260.

USGS (United States Geological Survey). 2015. How Much Water Is There on, in, and Above the Earth? https://water.usgs.gov/edu/earthhowmuch.html (accessed May 15, 2015).

Chapter 2

Planning Policies

Reliable water supplies are vital to the growth, development, and long-term prosperity of any community. Water supply reliability depends directly on planning and foresight in securing and maintaining water supply sources. To set the foundation for a solid water supply plan, water managers should consider how previous planning efforts, regional or statewide plans, and other related planning work can be integrated into a local water supply plan. A variety of drivers can affect when and how a water supply is developed.

These factors should be considered as part of initial activities in developing a water resources plan, because they directly affect the outcome and long-term value of the plan. Upfront planning, in terms of establishing the audiences and means that the plan will be used to secure water supplies, is key to setting a firm foundation for the plan. Outreach to public stakeholders and integrating public input into the plan are key elements for community buy-in to identify the future water supply needs and priorities that are inherently unique to each community. These foundational aspects of water supply planning are discussed in this chapter.

LOCAL PLANNING HISTORY

Planning for future water needs is an ongoing process—not a one-time event. Yet many communities and utilities have never initiated a formal water resources or water supply plan. Some may have had a plan generated by their predecessors but may or may not have access to the basis or findings of those previous water-planning efforts. Other water providers have institutionalized a formal process with defined intervals between updates.

Those who have initiated planning processes often find that the foundational elements of a plan evolve over time. Underlying assumptions change, such as the local industrial and commercial bases, population dynamics, social values, or the priorities placed on water resources. Each underlying assumption can alter the criteria for evaluating and comparing supply sources. Such changes manifest themselves in both the demand and supply sides of the equation. For example, a source of supply that was not used because it was cost prohibitive becomes more attractive for use because less-expensive alternatives have become more costly or are no longer available. Each of these changes can be cause for updating a plan.

Drivers for Planning

Often there are multiple drivers for initiating and updating water resources plans, starting with the initial assumptions and projections from prior plans. Observed growth in number of customers and trends in demands tend to vary from projections. As a result, new plans need to update projections in light of a series of potential drivers. These drivers generally fall into one of the following categories, as detailed in Table 2-1:

- Changes in demand
- Changes in supply
- Raw and treated water quality
- Aging water supply infrastructure
- Implementation of conservation and reuse programs
- Instream flow, groundwater management, and source water quality requirements
- Effects of climate change
- Operational changes
- Capital improvement planning

In some cases, other governmental permitting or regulatory agencies may require that water resources plans be developed or updated on a recurring basis, or when triggered by certain permitting or regulatory actions. Because those situations are unique to local rules and requirements, they are not discussed in detail here, but the remaining items are more fully discussed below.

COORDINATION OF UTILITY-LEVEL PLANS WITH STATE AND REGIONAL WATER RESOURCES PLANS

Identification of Relevant Plans

Water resources planning is ultimately a local exercise with regional effects. The planning analyses and decisions are critically important to the future vitality of a community and must reflect unique local conditions. The characteristics of supply and demand unique to each locale—and the values and priorities of the end users of the water—can only be coordinated through local processes. However, there should be a strong and beneficial linkage between local and regional planning efforts.

Regional planning is often organized by political entity or boundaries (e.g., counties or joint powers authority), watershed, state level, or interstate basin level. First and foremost, regional planning serves a coordination role. Watersheds and aquifers transcend political boundaries and provider service areas, and in most cases, utilities share sources of supply with their neighbors to at least some degree.

Statewide and regional planning spans those boundaries and allows for coordinating plans between local entities. Such planning efforts can help identify situations where multiple providers are planning to use the same source in the future—and thereby head off conflicts that might otherwise occur between competing users and uses. Statewide and regional plans can also serve as a vital information resource to local planners. By providing the higher-altitude view—albeit with less detail than local plans—statewide and regional plans can be used by local planners as a resource for surface water and groundwater physical and legal availability, current and projected demands on supply sources, and projections of potential future shortages in terms of magnitude and frequency.

Table 2-1 Summary of drivers for integrated resource planning

Drivers for Planning	Primary Causes	Description of Drivers
Changes in demand	General	Water resources plans are founded on projections of future demand. Annual demands drive decisions and investments in water supply sources, while peak-day demands can dictate the sizing of water diversion, transmission, storage, treatment, and distribution infrastructure to match customers' daily and seasonal use patterns.
	Growth in service area	Growth in demand driven by population and employment growth is a fundamental element of traditional water supply planning. Perhaps more than any other factor, anticipated growth in a provider's customer base is a call to action to plan ahead to ensure continued reliable water service for all customers. Many providers choose to plan for anticipated demands 50 years or more into the future to secure long-term reliable supplies. In some cases, obtaining permits and approvals for implementing water supply projects can take decades.
	New or expanded service areas	From time to time, changes in a provider's service area boundaries may occur, driven by land-use planning decisions, annexations, or intergovernmental agreements. The newly included service areas may be fully built out and developed, partially built out, redeveloped, or undeveloped lands. Portions of the new service areas may also be on domestic wells—not connected to any provider's treated water distribution network—and may be candidates for conversion to water service from the provider. "Wholesale" water transactions, where one provider sells water to another provider on an ongoing, intermittent, or emergency basis, can also directly affect the planning of both the seller's and the buyer's water supply portfolios. The buyer may purchase all or a portion of its water needs as raw or treated water from one or more sellers. Benefit–cost analyses for both the seller and the buyer should be conducted before entering into such agreements, but often there are economies of scale and operational benefits to creating regionalized systems involving partnerships between two or more communities, districts, or water providers. Each of these factors affects the amount of water needed and the timing and phasing for its implementation. Adding service areas that are already developed and served by an existing water system can add demand to a provider's supply system and treatment and distribution infrastructure gradually or literally overnight—requiring inter-local agreements, adequate planning, and the capacity to be identified and secured in advance.
	Uncertainty in long-range planning	There are many challenges in projecting service area demands far into the future. One particularly challenging aspect of long-range planning is the availability of population and employment projections that reach as far into the future as the water supply plan seeks to project. For example, many local land-use and population planning documents only project conditions 10 or 20 years out. Extrapolating from those projections for water supply planning, while often necessary, causes further uncertainty in the accuracy of long-range water demands.
	Variation from previous projections	In water planning, projections will almost certainly be more accurate in the first 5 years than they will be 50 years out. The assumptions and data used to make projections are often based in system-specific empirical data, observations, and trends. Actual demands almost always vary from projections—the question, then, is to what degree they might diverge. Water utilities can expect to see changes in annual and/or peak demands relative to previously projected values. This is often driven by deviation from the foundational assumptions in previous plans, such as • service area population and employment growth, • per capita demand, • new or modified inter-provider arrangements to buy or sell water, and • climate change. Per capita demands can be affected by changes in the nature of demands in the service area (e.g., a new major industrial water demand), the effectiveness and pace of adopting conservation programs or drought management measures, and short- and long-term trends in climate. Because these factors can trend upward or downward over time—and in many cases are difficult to predict with accuracy more than 5 or 10 years in advance—regular updates to the water supply plan are valuable in refining projections of per capita demands and the resulting implications on the source of supply investment and operational decisions.

Table continues next page

Table 2-1 Summary of drivers for integrated resource planning (continued)

Drivers for Planning	Primary Causes	Description of Drivers
Changes in supply	General	The traditional norm is that new or expanded sources of supply are prompted by growth in demand—and successful planning provides sufficient time for identifying and comparing new source options and to demand capacity measures, decision making, securing new sources for long-term use, and implementing the infrastructure needed to deliver reliable supplies. However, this may not hold true because needs for power, competing utilities, or industrial uses may quickly limit the access a utility has to its planned-for long-term water supplies.
	Climate change and surface water supply availability	Changes in supply availability or reliability can prompt a new or updated water supply plan. Adapting to climate change is at the forefront of many water providers' concerns, with the potential for significant changes in yields, water quality, and the timing of supply availability. In many cases, supply availability projections are indicating trends contrary to traditional assumptions, and historical data and observations. Projections in some areas call for wetter or drier conditions, more intensive precipitation events, or changes in snowpack and snowmelt patterns that can alter the supply picture and management approach. The Water Utility Climate Alliance (WUCA) produced a white paper in 2010 that discussed the use of multiple climate scenarios in water utility planning (WUCA 2010). The US Environmental Protection Agency (USEPA) developed a range of adaptive planning tools and resources for water providers' use in planning for climate change, embodied in the Climate Ready Water Utilities initiative (USEPA 2012a). A joint publication by the American Water Works Association (AWWA) and the International Water Association also provides insights into the future of water management under evolving climate conditions (Smith, Howe, and Henderson 2010). AWWA also established a policy regarding climate change (AWWA 2014a). Climate can affect water supplies in a variety of ways. In the northwestern United States, glacial melt continues to decrease stored water supplies, while in low-lying coastal areas, rising sea levels may encourage salt water intrusion into coastal aquifers. The variability and intensity of storms are changing in other locations. Utilities relying on snowpack for seasonal storage should assess the effects of projected changes in snowpack and runoff patterns. Long-term planning should include understanding the effects of climate as they affect water supply availability.
	Groundwater yield	The US Geological Survey Circular 1323 (USGS 2008) identifies much of the United States as having overdrafted aquifers. Many of these aquifers will not recharge anytime in the relative future. Much of the United States continues to pump these aquifers despite suggestions that water supplies will no longer be physically or economically viable in the next 30–100 years. Many of these areas lack major surface supplies, so the loss of groundwater may create significant water supply challenges for affected groundwater utilities. Because many communities, especially smaller communities with limited resources, built their water supplies around a network of groundwater wells, the challenge is daunting. Ultimately, in the long term, a supply that is not used sustainably will eventually become unreliable. Thus, an understanding of the supply/demand balance for any given aquifer, and the potential for regulatory limitations, can prompt the need to initiate or adapt water supply planning activities. Many providers find that no single source of supply is optimal. Groundwater is typically much more reliable than surface water supplies in times of drought, but it may be possible to overdraft aquifers beyond their rate of recharge. Surface water is renewable but subject to hydrologic variability. Both groundwater and surface water supplies can be affected by others' use of the same source of supply. In some cases, this has caused utilities to not only modify their plans for future supply additions but to initiate plans to transition to a more diversified water supply to develop a sustainable and reliable water supply network. All utilities should have sufficient capacity and redundancy to meet at least the basic indoor domestic water needs under any foreseeable supply conditions. All utilities should have drought response plans.
	Environmental flows	Environmental concerns related to river flows or water quality can also alter the supply picture. Minimum flow requirements may change or be newly promulgated. Impacts to endangered or threatened species could affect supply availability. Furthermore, surface water diversions can affect a variety of riverine water quality parameters, such as salinity and dissolved oxygen. New or existing water quality requirements can also affect a water provider's supply portfolio.

Table continues next page

Table 2-1 Summary of drivers for integrated resource planning (continued)

Drivers for Planning	Primary Causes	Description of Drivers
Raw and treated water quality	General	Water quality is integral to any water planning exercise and can directly or indirectly affect the decisions made in planning future water supplies. Much like water quantity—where changes in either demand or supply assumptions can affect the need for, and timing of, planning and implementation—water quality can be affected by both the end users' needs and the source of supply. There is a direct link between drinking water regulations, water quality, and public health protection. Additional information is provided in chapter 6 of this manual.
	Drinking water regulations	Ever-evolving drinking water regulations are perhaps the most significant driver for treated water quality. Faced with limits on new parameters or more stringent limits on parameters that are already regulated, water providers must balance the costs of treatment versus the choice of supply sources and the raw water quality characteristics inherent to each. For example, when the USEPA set new national standards for arsenic in drinking water in 2001, many communities faced a decision between implementing treatment for arsenic removal and abandoning existing water supply wells. Abandoning wells, in turn, would prompt the need to accelerate implementing alternative sources.
	Aesthetics of treated water	Users' preferences and provider goals for treated water aesthetics play a role in supply planning. While virtually any goal can be achieved through selecting the right treatment technologies, the costs of achieving those goals will vary with the quality of the source of raw water supplies.
	Cost of various sources	The quality of raw water supplies directly affects the costs needed to achieve regulatory compliance and user preferences for treated water quality. This is an important consideration when comparing the costs and operational implications of one source to another. In the United States, the Safe Drinking Water Act requires compliance with public health standards regardless of the costs. This creates significant challenges in some communities.
Aging water supply infrastructure	General	Water supply systems are a complex network of diversion, pumping, transmission, storage, treatment, and distribution infrastructure. Elements include structural components; electrical, instrumentation, and controls systems; piping; and equipment—each of which has a unique service life that depends on its use and ambient conditions.
	Incremental funding and replacement	Water systems are typically expanded and updated incrementally over time as needs and funding sources dictate. This translates to a situation where each year, different components of water supply systems are coming to the end of their useful service life. Decisions on rehabilitating and replacing existing infrastructure should be weighed in the context of considering alternative sources of supply or alternative means of diverting, conveying, storing, and treating the existing sources of supply.
	Maintenance	Preventive maintenance is key to reduce the potential for infrastructure and equipment failures, and the associated potential for supply interruption. Now more than ever, water providers are turning to asset management strategies and programs to forecast rehabilitation and replacement needs using data-driven analyses to target the financial needs without prematurely replacing equipment or relying excessively on reactive repairs.
	System reliability	Aging water supply infrastructure must be considered when assessing the reliability of water supply systems. Simply assuming that today's sources of supply can be used indefinitely without major investment within the planning period is unrealistic. Decisions integral to the water supply planning process include assessing the condition of existing infrastructure, considering its anticipated service life, and integrating these decisions with a robust asset management plan.
Implementation of conservation and reuse programs	General	Managing a water portfolio is a dynamic process, reacting to daily, seasonal, and year-to-year variability in supplies and demands. Some of this variability is driven by programs under the control of water supply planners, while others are driven by factors well outside of water supply planners' control. The components of the urban water cycle are increasingly recognized by planners and the public as being interconnected—and highly interdependent. Water conservation programs and water reuse systems affect this balance via two very distinct pathways.

Table continues next page

Table 2-1 Summary of drivers for integrated resource planning (continued)

Drivers for Planning	Primary Causes	Description of Drivers
	Water conservation	Water conservation is often considered in terms of reduction of water use or waste, and increased efficiency—using less water to achieve a given purpose. Often considered a component of a water *supply* portfolio, conservation *reduces demands* and can affect both annual and peak-day demands. Reducing demands has the potential to affect utility revenues, but conservation typically increases gradually over time, providing time for utilities to modify rate structures if needed.
	Water reuse	Water reuse does not reduce demands. Rather, it replaces traditional raw or treated water supplies with an alternative source of supply (typically highly treated water from a water reclamation facility) to help satisfy demands. Water reuse within a community can reduce or defer the need for developing additional new water supplies. This holds true whether the reuse system is directly or indirectly connected to beneficial uses, serving nonpotable uses, or supplying water for potable treatment and use. Water reuse decisions are often prompted by the economics of supply options or regulatory requirements. The economics can include the availability and reliability of the supply, the costs to implement alternative raw water supplies, and the economics of meeting increasingly stringent wastewater discharge permit conditions. Water reuse in upstream communities, in contrast, can affect the availability of surface water supplies for a downstream community, subject to water rights and permit constraints. To the extent that those factors drive additional water reuse, they also need to be included in water supply planning.
	Overall demand reductions	The degree to which a provider relies on conservation and reuse—and the degree and pace they are adopted and implemented in the community—can affect the water supply/demand balance. Part of a community's planning process might include an analysis of demand management measures, including their cost-effectiveness and water savings efficiency. Significant reductions in demand can defer the need for new water supplies, thus deferring capital expenditures, and operation and maintenance costs. Deviations from previous planning values in any of these factors affect the timing and amount of need for additional water supply solutions. One approach that many providers have found effective is the use of demand-based triggers for water supply capital projects, in tandem with the use of conservation and reuse programs. The timing of those demands and projects can be estimated based on demand projections, then updated over time—adjusting *when* the actions will be taken, while maintaining consistency in *what* actions will be taken to meet demands.
Instream flow, groundwater management, and source water quality requirements	General	For water managers overseeing portions, or all, of the urban water cycle, it is well understood that water quality standards drive nearly continuous investments and improvements in water reclamation facilities. At the same time, as users of water resources for potable supply, it is important to recognize the value that those water quality standards provide in terms of suitable source water quality for treatment and delivery to customers. In the water environment, the connection works both ways.
	Source water quality	While seeking source water quality for potable treatment and supply, the source of that supply and its quality will also affect the quality of the wastewater that is returned to the water reclamation facility for treatment and discharge. In some cases, standards for specific constituents (e.g., certain metals) may have discharge standards that are more stringent than those for drinking water. For example, metals standards for instream aquatic life beneficial uses may be an order of magnitude lower than the maximum contaminant level set by the US Safe Drinking Water Act standards. In that event, choices between sources of supply and operation of the water supply system can be driven in part by discharge standards at the water reclamation facility. These considerations should be screened as part of evaluating existing and potential future sources of supply—and investigated in more detail if warranted, based on the findings of the screening evaluation before making long-term decisions on water supply sources and their operation.
	Groundwater management	Voluntary or mandated restrictions on the rate of groundwater use can drive water supply planning efforts. For example, the 2014 Sustainable Groundwater Management Act in California requires the development of groundwater sustainability plans for overdrafted aquifers. The act is administered by the state on a groundwater basin basis in areas where there is no local or regional agency sustainably managing its groundwater levels.

Table continues next page

Table 2-1 Summary of drivers for integrated resource planning (continued)

Drivers for Planning	Primary Causes	Description of Drivers
	Instream flow requirements	Water planning by municipal water providers and water authorities has historically focused on the needs of their customers—water diverted or pumped, stored, treated, and conveyed through distribution systems to the end user. Today's planning often considers the implications of such diversions and uses on streamflows and nonconsumptive instream flow uses for recreational and environmental purposes. Those drivers—whether considering the effects of additional changes to instream flows or contemplating alternate water management methods—are key elements of an integrated water resources management approach to planning. Such considerations can affect when a new or updated plan is needed, what supply options and management methods are considered, and ultimately, the decisions that result in implementing new water supply, conservation, and reuse programs.
Operational changes	General	Managing a water supply system requires consideration of operational costs and implications.
	Operational cost impacts	Costs are driven by a range of factors, but only portions of those are within a water provider's direct control. Operational changes might consist of reversing the roles of sources used for peaking and baseloading. Those decisions can be founded on changes in operational costs (such as energy costs, capital costs, or labor costs) or on physical supply availability, or water rights, or permit constraints. Operational drivers should be fully characterized—and forecasted to the extent practical—to allow operational costs and preferences to be integrated into decisions between alternative source options and strategies for how those sources will be used on an annual and seasonal basis.
	Energy demands	A water provider with deep groundwater wells in an aquifer with limited natural recharge may be physically capable of pumping those wells to produce a given yield but may find that the available supply continues to diminish. Groundwater modeling techniques can be used to project future water-table levels and aquifer yields. Over time, in this situation, the model may predict a need to pump water from the aquifer at lower per-well yields. The increased energy cost to pump that water—with unit energy costs increasing at the same time—may make that source economically unattractive on a life-cycle cost basis, prompting a decision to accelerate the implementation of alternative sources even though the wells could continue to produce water. Conversely, pumping energy costs for a regional surface water supply project could escalate to a point where that source is less attractive than others. Oftentimes, planners are faced with a decision between locally available—but expensive to pump and treat because of aquifer levels or local water quality—and more distant regional supplies that have higher water quality but higher transmission costs.
Capital improvement planning	General	Water resources plans can also be initiated as part of a utility's regular cycle of capital improvement planning. Because it can take decades to obtain permit approvals and funding for major projects, water supply plans often look well beyond the typical time frame of a utility's typical capital improvement plan time horizon. Nevertheless, near-term costs can be estimated for inclusion in capital improvement plans, and longer-term cost ranges can be used to help the utility assess longer-term rate structures and system development fees.

Local planners should be aware of regional or statewide water planning efforts because they have been engaged in those planning processes. As a result, local planners understand the types of information made available by the state or regional process, and the constraints and limits of those processes.

Statewide planning varies widely from state to state. Some states have never undertaken a statewide planning effort. Some have initiated statewide planning processes in the last decade. Others have been conducting planning and periodic plan updates for decades. The US Army Corps of Engineers (USACE) conducted an assessment of statewide planning efforts, culminating in the 2010 report titled "Responding to National Water Resources Challenges" (USACE 2010). That report provides a list of lead state agencies for water planning in each state.

Regional efforts can be identified through a variety of avenues, such as professional associations, state and federal water agency contacts, and regional political organizations and venues.

Planning Data Resources and Tools

Previous plans can serve as a key foundation for updates. Professional and industry associations (e.g., AWWA, American Water Resources Association, WateReuse Association) also offer a wide range of resources—including this manual—for water supply planning. Related tools for developing plans and conducting analyses for water conservation programs, water reuse systems, and other facets of integrated water management are also available.

Water demand projections may be available from regional or statewide plans. Similarly, foundational data for demand projections, such as population and employment projections, may be available from sources such as city planning departments, the United States Census, state demographers, economic development organizations, transportation planning agencies, or state universities. These data may or may not be on the same scale as the local plan. For example, population or demand projections may be forecasted by others on a city, county, or watershed basis. Those boundaries—and the resulting projections—may not match with those of the local planner's service area. Many water providers' service areas do not directly coincide with other political or watershed boundaries. Even within a community, some of the residents may not be served by municipal or rural water systems (e.g., individual wells), or there may be service outside the community's boundaries. This may require using geographical information systems or other tools for manipulating the forecasts to reflect the local agency's planning boundaries. Local planning numbers from more detailed local studies may also be available for use as foundational data. State and regional planning should generally not supersede local plans but rather can serve as a resource for local plans and can be used for comparisons between areas by applying consistent methods and sources of data. Careful attention needs to be paid when combining plans, because population estimates may be higher or lower due to overlapping boundaries of the planning agencies or estimating methods for the population or employment. For example, for a census tract that spans across two water agency boundaries, both agencies would need to agree on the proportion of the population that belongs in each service area, especially if regional or state planning estimates are being used based on local plans.

Supply data may also be available from other sources. The US Geological Survey (USGS) maintains a nationwide network of streamflow gauging stations and groundwater-level monitoring wells, and makes those data publicly available through its online National Water Information System (USGS 2012). Other streamflow data from state or federal agencies may be available in the watershed of interest. Groundwater aquifer data may be available from the USGS, state sources, and local or site-specific studies.

Water quality data should also be sought as part of planning and anticipating potential water treatment needs, using federal sources such as the nationwide Storage and Retrieval of US Waterways Parametric Data (STORET) database.

Coordination of Results

Active engagement by local planners in regional, watershed, and statewide planning processes is key to the success of those efforts in terms of coordinating water supply planning and heading off conflicts between competing uses before they occur. Participation by local planners in such processes is important, regardless of what stage of their planning they may be conducting or implementing at the time.

INITIAL SCOPING

Clearly scoping the effort and end products for a water resources plan will focus the planning team's efforts from the outset, resulting in a targeted plan that meets established objectives and is most useful to the target audiences and users of the plan. Initial scoping includes taking into consideration who the primary and secondary "audiences" or users of the plan will be; clearly setting goals, objectives, and outcomes for the plan; assessing what is and is not within the scope of the planning effort; and setting a timeline for planning.

Audiences

Each water resources plan should focus on its users and uses as part of early efforts to organize and "plan the plan." By understanding how the plan will be used and when it will be updated, each of the plan's elements can be targeted at meeting those goals. Examples of the factors that should be considered as part of early planning efforts are the following:

- Who has authority to approve or modify the plan?
- Who will be responsible for implementing the plan?
- Who might be affected by the plan—both within and outside the local provider's organization, and over the near- and long-term time frames as it is implemented?
- Who will monitor implementation progress and adapt the plan as implementation proceeds?
- Who will be responsible for updating the plan, and when will that update be initiated?
- How will customers benefit from the plan?

If the answers to these questions are not immediately clear, the planning team should work with internal and external stakeholders to resolve and clarify the role of the plan in addressing future water needs and water management strategies. Ultimately, the plan has limited value if the stakeholders, especially the users, do not buy into and implement the proposed plan.

Priority Goals, Objectives, and Outcomes

Initial planning activities should focus on the end goals in terms of the priority outcomes from the planning process. What will the organization know at the end of the plan that it does not know now? What do planners and managers need to know to continue to have a safe and reliable water supply portfolio decades from now? What specific information and direction will the implementation team need to know to put implementation of the plan into motion? A major priority should include matching demand patterns with water supply options. The critical juncture for any water supply plan is when the current supply availability is exceeded by the projected demands. Securing a new supply is often a difficult task and will require significant financial investment and commitment on the part of the stakeholders.

Planning Boundaries—What Is On and Off the Table?

The planning team should take a deliberate approach to setting boundaries for planning, once the primary and secondary audiences are identified and the priority outcomes have been established. The following are examples of the types of questions that should be answered:

- What is the planning horizon (e.g., 50 years)?
- How much detail will be included in near-term years versus longer-term planning years?
- Will the plan consider meeting nonpotable demands through water reuse or raw water supply options?
- Is indirect potable reuse (supply augmentation with highly treated water from a water reclamation facility) a potential consideration? Is direct potable reuse a potential consideration? Why or why not?
- Should the plan include developing a conservation plan? Is there already a conservation program whose goals should be assumed as part of basic demand projection work? Does the conservation plan contemplate efficient, cost-effective measures that could extend supplies and/or reduce capital outlay?
- Are there any specific supply sources that must be included or are definitely excluded? Why?

By clearly delineating the uses and boundaries of the plan—envisioning the end product and its users before developing the plan—planners can focus their efforts on the elements that most directly benefit the intended audience and users of the plan. This will likely include both internal stakeholders such as a city council, authority board of directors, and the provider's staff charged with implementing the plan, and external stakeholders such as the provider's customers, neighboring communities, and water-related interest groups.

Timeline for Planning

When considering sources of supply, many plans take a 50-year or longer outlook. It is often difficult for stakeholders, and especially elected officials, to embrace long-term plans. They often ask, "Why so long, when the accuracy will be diminished the further out demands and supplies are forecasted?" The answer is that developing new supplies may take 15 to 20 years, and understanding how growth might occur in the 50- to 100-year period will provide justification for acquiring appropriate allocations of those new supplies. Most stakeholders understand the importance of a permanent—and permanently reliable—water supply system. Too often, permitting and regulatory processes create significant time delays in the implementation process. Major projects, such as new reservoir storage, can trigger a decades-long regulatory process. Thus, it is also important to project long-term needs and phasing schedules to fully understand when supplies might be at risk for not meeting demands. As competition grows for finite resources, so grows the potential for conflicts and shortages. A solid look ahead in light of shared use of water resources can head off problems before they materialize. It is also important to project longer time horizons to take into account incremental changes (such as loss of storage capacity) or the increase of frequency and intensity of extreme climate change events, which are small shifts on a near-term basis, but when aggregated over a longer time horizon can have a significant effect on planning assumptions for the permanence of supplies and planning outcomes.

PROCESS AND ROAD MAP FOR PLAN DEVELOPMENT

AWWA Policy

Water rights and allocation of water supplies is fundamental to any water supply planning process. Water rights are described in detail in chapter 3 of this manual. AWWA has developed the following policy regarding developing and managing water resources (AWWA 2014b):

> *The American Water Works Association (AWWA) supports and promotes the integrated use of adaptive water resources planning processes and effective management practices to provide for reliable, sustainable, and cost-effective supplies of appropriate quality water for customers.*
>
> *Freshwater sources are increasingly at risk from a variety of natural and human-induced stressors, including population increase, climate change, land-use changes, and pollution. Additionally, water system reliability is increasingly at risk due to aging infrastructure that is in need of replacement or renewal.*
>
> *In order to assure current and future water supply reliability, it is essential that water managers use iterative and adaptive long-range planning processes that encourage consideration of local and regional water supply options, including demand management. These efforts should be comprehensive in approach, include the participation of local and regional partners, and explicitly evaluate water resource conditions under a variety of scenarios. Additionally, water managers should routinely review their water management practices and consider all opportunities for incorporating operational efficiencies that result in water supply, water quality and cost benefits.*
>
> *Water agencies should consider all water supply options that are potentially viable. Surface and groundwater storage should be considered, when and where feasible, as an effective approach to managing and augmenting local water supplies. In addition to freshwater sources, potable options include treating brackish groundwater, seawater or reclaimed water to drinking water standards. Nonpotable supply options such as reclaimed/recycled water, decentralized rainwater and greywater collection systems, along with stormwater capture, are all effective options for reducing demands on potable supplies and should be considered in the planning process. These options should be evaluated with consideration of life-cycle costs at local and regional levels to optimize the long-term value of investments in utility infrastructure.*

Planning Team

The planning team should include a core group of water professionals at its foundation. In smaller communities, this may include one or two lead staff, while larger utilities may have several key staff from their management, planning, operations, conservation, or engineering groups. Depending on the level of detail and effort targeted for the plan, and the provider's in-house staff availability and qualifications, the provider may elect to engage the services of consultants, university staff, or other supplemental resources.

The planning team should include, or have direct access to, a broad set of viewpoints. For example, the planning team should actively seek and address input from operational, financial, engineering, management, and council or board representatives. It should also reflect public input and involvement through a strategic effort of the utility's public information officer or consultant. Some providers also seek out advisory panels or steering committees to bring representative viewpoints for various interests in the community. In establishing the planning team, consideration should be given to including representatives from organizations that could affect—or be affected by—the outcome of the plan. Examples might include key permitting agencies, public interest groups, or other neighboring water providers.

Identifying and communicating each team member's specific role, responsibilities, and expectations in the planning process should be clearly established at the outset of the project.

Defining Timelines and Responsibilities

Organizational planning for developing the water supply plan should include documenting key timelines and the roles and responsibilities of each team member—both on an individual and organizational level. Identifying the critical path for plan development activities can be a useful tool in focusing the team's efforts. Often, maintaining a decision log, an action item tracking log, and similar project management tools can be helpful in establishing and maintaining a path toward timely completion of the plan. Sufficient time should be built into the schedule to obtain necessary internal team reviews, and for meaningful public involvement and input. Given the importance and long-term ramifications of the decisions made in a water supply plan, adequate time for deliberation and decision-making is crucial to a successful plan that sets a course for a long-term reliable water supply.

Organization of the Plan

No single outline exists that is best for all planning processes. The organization of the plan will in many ways depend on the goals and intended audiences for the plan.

That said, many plans include the following basic elements:

- Introduction and drivers for planning
- Basis of planning (demand projections, anticipated regulations, etc.)
- Water supply options
- Screening of viable options
- Developing and evaluating "portfolios" of supply
- Public involvement process
- Recommendations and capital improvement plan data

The process outlined in this AWWA manual provides detail on the development of each of these aspects of planning, with suggestions on organization and content for each.

Planning Process and Decision Making

Deliberate consideration of how decisions will be made—the criteria and process that will be used to compare and select alternatives—should be conducted well in advance. In many cases, participants and stakeholders will be much more willing to accept and constructively improve the results of a plan if they "buy in" to the methods that will be used in the planning process. Decision making can use simplified pro/con methods or more advanced multicriteria decision support methods. While each provider will have its own drivers for selecting planning process and decision making methods, using multicriteria decision support methods can be useful in documenting to other stakeholders (including city councils, boards, etc.) the rationale and basis for the decisions that are made in the planning process.

In some cases, internal and external stakeholders can participate in the decision making process. For example, when considering a seawater desalination plant for Florida's east coast, project team members held workshops with internal stakeholders to determine and define evaluation criteria. Once each criterion was defined, it was presented to the community for review and input. Additionally, internal and external stakeholders provided input on the evaluation criteria weighting. Multiple weighting scenarios were developed and applied to help make the decision makers' selection.

PUBLIC INVOLVEMENT

Public involvement has become an essential step in the planning process. Stakeholder and interest groups are more sophisticated in their communication skills and are more apt to understand and use the power of mass and social media to marshal political and voter forces.

Whether it is required by statute or simply a commitment on behalf of the utility, public involvement provides water managers with a broader community context as they consider demand, supply, regulatory, economic, environmental, and other technical aspects of identifying new water supplies. Public involvement is more than just a public meeting—it is a strategic process to engage stakeholders early in the planning process so that they may contribute to and influence decision making. An open, transparent decision-making process can lead to an informed, consensus-based decision that reflects community values and minimizes public opposition.

AWWA's policy statement on public involvement and customer communication states, "Opportunities for input and involvement are essential to public understanding and acceptance of utility programs and projects. The format and depth of involvement will vary according to individual suppliers, communities, and issues. Opportunities for involvement must, however, be meaningful, inclusive, and clearly linked to the decision-making process" (AWWA 2013).

Benefits of Effective Public Involvement in Resource Planning

Today's water suppliers must balance multiple, often competing organizational and political demands while also providing safe, high-quality drinking water to their customers. The water resources planning process seeks to frame resource decision making in terms of community values, as well as technical and cost criteria. Therefore, the planning process should be designed to not only solicit community input but also use that input in a meaningful way during decision making. There are numerous benefits of meaningful public involvement:

- Aligning the utility's projects and plans with its constituents' long-term plans and priorities
- Listening to, and learning from, the public to develop and implement projects that incorporate community values and priorities
- Increasing public access to accurate information about current water quality and quantity, and other affected water resources, thereby increasing trust in the utility and providing a counterbalance to misinformation that may be communicated through other avenues
- Increasing the public's baseline understanding of water permitting and water quality regulations that have dramatically increased the need for more costly advanced treatment technologies
- Increasing public support for balancing development of limited water resources and increasing need for water use efficiency and emphasis on environmental stewardship
- Demonstrating the utility's commitment to efficient and effective use of public dollars
- Providing more and better information that is otherwise unavailable
- Increasing elected officials' comfort level with complex decisions about long-term investments in a community's future by building a public mandate for sustainable, high-quality water
- Reducing the potential for protracted win/lose battles with project opponents, lawsuits, and the odds that needed permits may be denied

Integrating water resources planning and public participation can add credibility to a project and can improve public trust in government/institutions. Public involvement can help utilities garner the support necessary to move a project forward. Furthermore, public involvement can help projects succeed by reducing the risk of failure, ensuring compliance with regulatory agencies, and gaining acceptance with the public and elected officials.

Planning for Public Involvement

The International Association for Public Participation (IAP2) states that public participation means to involve those who are affected by a decision in the decision-making process (IAP2 2012). Integrating public participation into the planning process requires a concerted, strategic effort to identify and engage stakeholders to build understanding and incorporate public input into decision making.

Stakeholder groups include many parties; consideration of local and regional stakeholders is imperative. Such groups might include local economic development interests, business leaders, ratepayers, environmental groups, and regulatory agencies, while regional groups might include upstream and downstream users, agricultural interests, industries, and other government agencies. As a part of the process, some effort should be undertaken toward a risk communications framework, if there are perceived health risks, potential controversy, or low levels of trust.

Truly involving the public in identifying, prioritizing, and assessing the value of resources will require a strategically planned process for reaching interested members of the public, educating them on the issues, and providing a meaningful way for them to express their opinions and influence the decision making, enabling them to be effective participants in the decisions that are made.

Planning for public involvement should be integrated with the overall water resources planning process, beginning in the conceptual planning phase. The decision to include the stakeholders requires that an honest assessment of water supply alternatives and benefits will be the basis for the planning effort, and that the public's input will have a bearing on the final decision-making process. Assuming that the planning process will be collaborative, the development of an effective public involvement program should begin with careful, thoughtful planning. Essential elements to successful planning include the following:

- Defining the project and assessing potential risks to the successful completion of the project
- Assessing the "owner's" (utility or decision-making entity) readiness for public involvement and ensuring that adequate budget and staff are allocated to the effort
- Conducting a preliminary assessment of potential stakeholders and their concerns and expectations for involvement
- Identifying techniques for informing and involving stakeholders
- Developing, implementing, and monitoring the public involvement work plan

Special planning considerations. Public involvement becomes a regulatory requirement under certain conditions. Some states require review of projects affecting large areas. The National Environmental Policy Act (NEPA) and National Historic Preservation Act (NHPA) require public involvement when there is the potential to impact environmental, historic, or cultural resources where a federally permitted project is being considered.

Additionally, USEPA has an environmental justice mandate that extends to all of the agency's work, including permitting facilities, awarding grants, and reviewing proposed actions by federal agencies. USEPA's website defines *environmental justice* as "the fair treatment and meaningful involvement of all people regardless of race, color, national origin, or income with respect to the development, implementation, and enforcement of environmental laws, regulations, and policies" (USEPA 2012b).

If a proposed project or planning effort affects a recognized Native American tribe, consultation is taken to another level. Federally recognized tribes are recognized under law as sovereign nations. Within the United States, there are about 570 federally recognized Native American tribes or governments. NEPA and NHPA require that consultation and collaboration with affected tribes be conducted in a formal government-to-government manner. Additional information on public involvement and NEPA, NHPA, environmental justice, and Native American tribes can be found online at the following links:

- USEPA National Environmental Policy Act—www.epa.gov/compliance/nepa/
- National Historic Preservation Act—www.achp.gov/about.html
- Environmental justice—www.epa.gov/environmentaljustice/
- Native American tribes—www.epa.gov/sites/production/files/2015-03/documents/ips-consultation-guide_0.pdf

A team approach. Effective public information planning and implementation requires input from, and coordination with, the water resources planning team. Ideally, the public involvement task leader should be a key member of the project management team to ensure collaboration, alignment with the overall planning process, and improved communication between stakeholders and the project team or utility. This also ensures that stakeholders are involved at strategic opportunities within the project schedule, so that input can influence decisions before they are made.

The project team members ranging from senior project management to planning, permitting, funding, and design can play a role in public involvement planning and implementation. During the public involvement planning stage, it is helpful to identify an internal project champion. The champion is the designated senior-level manager who is visible in the community, has access to political decision makers, and has a passion for the project to succeed.

Champions have a big-picture view and are able to accommodate temporary setbacks; they consistently communicate both internally and externally about the progress being made; and they are able to maintain project momentum during controversy. External champions are also essential project team members. They are typically allies in the community who act as a conduit about public sentiments. They often serve as contacts and symbols for elected officials, especially during controversial projects, because they demonstrate knowledge of community values, understand project priorities, and illustrate sensitivity to local politics. They can be important local opinion leaders who speak on the team's behalf.

Laying the foundation. The public involvement planning process starts with the water resources planning team. Input from the team is necessary to lay a foundation for the overall process, help define goals, and manage expectations. Key foundation elements include the following points.

Framing the project. Determining the specific problems that will be solved, and identifying the boundaries of those problems, helps to frame the project in terms that both internal and external interests can understand. Facts that establish the need and scope of the process are clearly defined, and assumptions used in the process are identified. Once these boundaries are understood, it will be easier to establish achievable goals and expectations.

Distinguishing facts from assumptions. Water supply alternatives are more clearly defined when facts are distinguished from assumptions. Facts are limited to what is known about an option: they can be readily agreed upon. Project facts may focus on water supply capacity or permitting constraints. Assumptions often form the basis for disagreement, so it is important to uncover differing opinions about data sources, uncertainties, regulatory projections, and planning horizons—first within the sponsoring organization and then among key members of the public. During this planning phase, facts and assumptions should be listed separately as distinct line items to ensure that all potential obstacles are anticipated.

Developing a need statement. One of the most important components of the public involvement planning process is developing a clearly defined water supply need statement. The need statement consists of a few sentences describing why the utility must resolve particular problems. Developing a concise, easy-to-understand need statement will foster internal alignment among project team members but also clearly convey the need with staff, the media, elected officials, and the public. Sharing the need statement outside of the utility will foster internal alignment, gain the attention of the public, and lay the groundwork for consensus-based solutions. Sometimes needs can be defined jointly between the utility and the stakeholders.

Planners should resist the temptation to develop solutions to problems too early in the process. It is natural to want to present solutions when describing a problem to the public rather than just communicating about the problem. Instead of bolstering credibility, this position can alienate the public, forcing the utility to try to sell the project, while pushing the public into positions of support or opposition. At this stage, it is essential that a public forum be established for discussing problems. In this way, managers can resolve public concerns rather than brokering positions on a project. In the field of mediation, this approach is referred to as "interest-based negotiation" rather than "position-based negotiation."

Determine the Process and Opportunities for Public Involvement

For public involvement to be meaningful to both the participants and the decision makers, the public's input should feed into the process at logical points. Once the opportunities for public involvement are identified, planning can take place for appropriate activities or venues.

To gain public confidence, utilities look for external champions among stakeholder groups. This is best accomplished by identifying interested members of the public who have the knowledge and background to understand most of the full range of decisions that must be made during the water resources planning process.

The following questions may be helpful in identifying when the public should be involved:

- Do opportunities exist for informing the public about the issues and challenges?
- Do project alternatives exist? If so, how can the public participate in developing or ranking alternatives?
- Do opportunities exist for public input into selection criteria or criteria weighting that will be used to rank alternatives?
- Does an opportunity exist for the public to participate in some aspect of water supply development, particularly location of facilities, site configuration, aesthetics, and timing?
- How much could and/or should alternative strategies, such as water conservation and water reuse, contribute to meeting future demands?
- Are there opportunities to create other community benefits, such as parks or environmental centers, related to this project?

Determine Public Involvement Priorities and Develop Budget

Public involvement is typically linked with project implementation. However, budgeting for public involvement activities should be aligned with each phase of water resources planning activities. This will ensure that resources and staff are available during the planning process. Implementation of a public involvement plan is typically front-end loaded with the bulk of the planning and public involvement activities taking place early in the water resources planning process.

Planners should begin by focusing on the overall plan, then considering each project and any special project-specific needs. Project issues raised in the past by customers, activists, elected officials, other communities, or the media should be factored into the process.

Securing the resources necessary to implement meaningful public involvement programs is a fundamental challenge. Because staff and financial resources are limited, the focus should target future projects with the greatest risk of opposition, delay, or failure. Examining future projects with public involvement will prompt a series of questions about the nature of the challenges a project might face and will assist in developing a strategy on how to allocate resources to each water resources project.

Assess risks and priority for each resource being evaluated. Project risk is the potential impact on the organization for the failure to develop or complete the project or program. Types of risk to water resource projects may not be inherently clear. Not developing a plan at all, or failure to develop a plan in a timely fashion, can be risky in terms of lost opportunities to secure a water supply option, a delay in the supply coming online that may lead to local moratoriums on development, or the inability to secure supplies in the face of a sudden change to, or loss of, a current supply that creates large, unanticipated financial losses. The investment in public involvement should be proportional to the risk posed to the organization, although with even a low-risk project, some minimal level of public involvement is always needed. Public involvement should not be viewed as an insurance policy against project failure. However, projects that include high risks or significant controversy should include robust public involvement programs. When assessing a project's risk, these variables should be considered:

- Consequences of not developing a feasible water resources plan
- Risks that eliminate water supply options
- Effects that the lack of public involvement may have on internal and external plan approval

After assessing risk, focus can be shifted to resources and projects that are more controversial than others, those that require public support, and whose success is key to the water resources plan as a whole.

Assign Responsibilities

To ensure a credible and sustainable public involvement plan, it is important to assign responsibility for individual public involvement activities, determine who has lead responsibility, and what support will be needed to provide the appropriate technical input, logistical support, and production assistance for public involvement materials. A well-planned public involvement initiative will include both labor and expense budgets for each task. The number of professional hours for each person and all expenses should be budgeted for each task. Senior management should evaluate the workload and determine whether outside consultants will be needed to provide staff support. Consultants can help balance workload without the time or expense of hiring full-time staff, and they offer expertise that may not be available in-house. They usually cannot substitute for staff, because internal staff typically has a far deeper understanding of the system, its ratepayers, and its challenges than consultants.

A strategic planning process will ensure that all alternatives have the appropriate level of public involvement, can help the utility garner internal and external support for public involvement, and will guide the development of a strategic, implementable public involvement work plan.

Prepare the Public Involvement Work Plan

After determining the elements of the planning process that will need to interface with public involvement activities, a work plan should be prepared. The work plan will outline specific tasks to be performed in concert with the water resources planning process. These activities include

- identifying stakeholders,
- identifying public involvement techniques,
- considering a professional facilitator, and
- matching public involvement activities with water resources planning elements.

Identify stakeholders. The public is not one group but many groups or individuals—each with different viewpoints about the utility, its problems, and potential solutions. These interested groups or individuals are called "stakeholders" because they have a stake in the outcome of the decision. It is important to remember that stakeholders are not limited to groups or individuals in favor of the water resources planning process.

Stakeholders often include opposition groups and vocal detractors. The following questions will help identify key stakeholders:

- Who has information?
- Who is likely to be concerned about the problem?
- Who will benefit by potential solutions?
- Who will be affected by potential solutions?
- Whose support is needed to implement the plan?

Project managers often think they know and understand stakeholder concerns, but these perceptions are sometimes wrong or have changed over time. To ensure that stakeholder views are clearly understood, one-on-one interviews should be scheduled with the people who are likely to shape opinions about the project. Interviews with these opinion leaders often correct assumptions about stakeholder opinion and provide valuable insights. These interviews are an opportunity to measure the credibility of the organization, further expand the network of project stakeholders, plan for information dissemination, and identify potential avenues for stakeholder input. To conduct a successful interview, a list of questions can be prepared that are designed to achieve the following objectives:

- Understand a group or individual's specific concerns about a particular problem or set of solutions.
- Determine which interests will be satisfied with receiving timely, appropriate information and which interests will seek to influence the decision-making process.
- Identify other stakeholders to interview.
- Determine if any of the stakeholder values are contradictory.
- Identify where and how interested groups or individuals are likely to find their information.

Stakeholders will have varying levels of interest and expectations on any given project. Typically they will fall into one of three groups:

1. **Group A** expects to be informed but not involved.
2. **Group B** expects to be able to exchange ideas with the project team.
3. **Group C** expects to be able to influence the outcome of the project.

Understanding the stakeholders and their level of expectations will help in determining which techniques will work best and how to link public involvement activities to project tasks. Table 2-2 is an example of how to organize information gained from stakeholder interviews.

Identify techniques. When selecting public involvement techniques, project needs must be linked with stakeholder expectations, the owner organization's culture, and the time and money available to devote to public involvement. Some techniques are suitable for informing the public, others are better suited to consulting with stakeholders, while others enable stakeholders to participate in some aspect of project decision making (see IAP2 Spectrum of Public Participation, IAP2 2012).

Internet-based tools, websites, and social media facilitate information and idea sharing and can play a part in any public involvement effort. Providing information via the Internet is expected in the current environment, but care should be taken when deciding the best method for exchanging information with stakeholders. Technology-based

Table 2-2 Organization of information from stakeholder interviews

Stakeholder	Issues and Concerns	Expectations for Involvement (A, B, or C)
Resident	Cost and effect on property values	A + B
Local elected official	Effect on constituents, effects to water rates	B + C
Environmental	Compliance with regulations, environmental impacts	B + C

techniques can supplement traditional efforts to increase public participation but cannot replace the intangible benefits of meeting face-to-face with the community.

Water resources projects most likely will use some techniques from all three categories. Efforts should focus on the most critical tasks associated with the water supply alternatives. A few common techniques are listed in Table 2-3.

Consider a professional facilitator. A professional facilitator can play a vital role when involving or collaborating with stakeholders in the decision-making process. A facilitator is a neutral third party that develops and implements a process to help a group provide focused input, reach consensus, or make decisions. Another definition is a person whose selection is acceptable to all the members of the group, who is substantively neutral, and who has no substantive decision-making authority but is someone who diagnoses and intervenes to help a group improve how it identifies and solves problems, and makes decisions to increase the group's effectiveness (Schwartz 2002).

A facilitator is particularly helpful when dealing with larger groups, such as in a workshop setting. A professional facilitator can help run the meeting or workshop, help develop ground rules, call on participants, manage the agenda, and more. More importantly, a professional facilitator's role is to understand the project owner's goals and develop and implement a framework for effective input to help achieve that goal.

An effective facilitator ensures that everyone is involved in the process and encourages open discussion. Because the facilitator is also impartial, this means he or she can help groups name problems and identify options, help avoid or resolve disputes, and work to maximize joint gains (Field 2010).

Match public involvement needs and opportunities with each planning process element. Identifying opportunities for incorporating public involvement includes combining what is known—about stakeholder interests and expectations for involvement—with what can be done to include the stakeholders in the decision-making process. One way to identify strategies for linking stakeholders to a project is to evaluate the project's task schedule.

A task schedule is a common project management tool often developed as a Gantt chart using Microsoft Project, Oracle's Primavera, or some other type of project planning software. Project management software illustrates the relationship of one task element to another over time. This is the project manager's road map and the key to understanding how to incorporate public interests into the decision-making process. Public input can only help shape the project if it is obtained before key decision points, so understanding the schedule is vital.

To enhance support for, and implementation of, a public involvement work plan, all public involvement activities should be displayed graphically on the project schedule. The goal is to link public involvement tactics with the other project tasks so that public involvement becomes an integral part of the project delivery process.

Understanding the relationship of the project tasks with the concerns of stakeholders is critical to the success of a public involvement program. An effective public involvement program can be crafted once the stakeholder concerns are linked to specific project decisions. The information gathered from stakeholder contact will help link the project tasks with public concerns.

Table 2-3 Informing the public

Getting Information to the Public	Getting Information From the Public	Exchanging Information With the Public
Internet (email, websites, social media, blogs, microblogs, etc.)	Internet (email, websites, social media, microblogs, etc.)	Internet (social media, microblogs, etc.)
Direct mail/bill stuffers	Questionnaires or surveys (telephone, Internet, direct mail)	Information line telephone number
Newspaper advertisements	Focus groups	Storefront, trailer, street kiosk
Door hangers	Small group meetings/interviews	Talk shows
Newsletters	Informational workshops	Speakers bureau
Displays	Public comment period	Door-to-door canvassing
Local media via news releases	Agenda items at local public meetings (councils, boards, or commissions)	Advisory groups
Articles in existing publications		Public meetings (open houses, presentation/Q&A, workshops)
Legal notices		Telephone town hall meetings
Cable TV		Web-based meetings

One vitally important step in the public involvement process is gaining and utilizing feedback from stakeholders, then conveying how that feedback was used. If stakeholders are uninformed how their input was used, it can harm credibility and confidence. Communicating back to stakeholders throughout a public involvement process is required to realize the benefits of early and inclusive public participation.

Implement and Monitor the Plan

Public involvement, like any other project component, requires focused, strategic implementation of the work plan. If the organization already has established project management processes, they should be used. While it may be necessary to modify the processes to address both the needs of the public involvement program and technical aspects of the project, existing tools are a good starting point.

Monitor the project. Predicting and managing project changes involves monitoring the status of a public involvement work plan almost daily. Regular communication with project team members and advisory bodies facilitates this process. During the monitoring phase of plan implementation, the project manager must be aware of the interrelationships between the project's public involvement activities and schedule of technical tasks to ensure a consistent workflow. Understanding the sequencing of tasks is also important because the plan must build on the successful completion of a preceding task before the next one can be implemented.

Active monitoring of the project will indicate where the public involvement program is working and where it needs improvement. The plan should be assessed to determine how effectively the individual tactics are meeting the intended goals. Assessments should focus on the most critical issues or stakeholder groups, because these have the potential to stop or significantly delay the planning process. The means of making these assessments include the following:

- Tracking social media for opinions, volume, and other parameters
- Asking key stakeholders their opinions
- Conducting public surveys to measure change in attitude or position

- Tracking media coverage
- Monitoring attendance at public meetings

When developing assessment tools, planners choose the methods that best meet the needs of the project. Their scientific defensibility may not be as important as their ability to gauge how targeted stakeholders feel. The feedback should be used to correct previous assumptions about opinions on proposed projects.

Manage change. Change on any project is inevitable. Public involvement programs must address public concerns and perceptions, which are often in flux and can be hard to predict. Even the most subtle change in public sentiment can cause a shift in the public involvement strategy, timing of program elements, and program goals. Public involvement programs must be flexible to respond to events such as election results, pending rules, evolving community issues, and changing stakeholder needs as awareness grows.

Because some changes in public opinion can be anticipated, it may be valuable to assign monitoring responsibility to the most probable risk-related changes. For example, the team member assigned to address political issues should also monitor local elections, candidates, and platforms. Ideally, the public involvement team should develop a change management plan that identifies potential changes in project parameters and public opinion, and methods for addressing them. Creating a flowchart that maps out hypothetical scenarios will help to display the plan. Sometimes public information committees have been used to guide needed mid-course corrections so that the goals of the public involvement plan are met.

Case Study: Norman, Oklahoma, 2060 Strategic Water Supply Plan

The City of Norman, Oklahoma, acting through the Norman Utilities Authority (NUA), provides a case study that exhibits many of the key drivers and elements for water supply planning. Located in central Oklahoma, on the southern edge of the Oklahoma City metropolitan area and home to the University of Oklahoma, Norman's service area, which included about 94,000 people in 2010, is projected to grow to a population of nearly 200,000 in 2060. NUA serves about 85 percent of the overall population of Norman, with the remainder on private domestic groundwater wells.

NUA draws on local groundwater supplies from the Garber-Wellington bedrock aquifer to meet about one-third of current demands, surface water from nearby Lake Thunderbird for about two-thirds of current demands, and intermittent use of treated water from an interconnection with the City of Oklahoma City to meet peak-day demands on an as-needed basis. The NUA *2040 Strategic Water Supply Plan* (SWSP) was developed in 2001, and updated in 2014 as the 2060 SWSP (NUA 2001, 2014).

Promulgation of significantly lower arsenic standards for potable water in the early 2000s led NUA to make an economic decision to shut down a significant number of its existing groundwater wells, rather than implement arsenic removal treatment. Potential future hexavalent chromium standards could prompt a similar future decision for NUA.

Drivers for initiating the 2060 SWSP included the following:

- Projections for significant growth in service area population, and associated increases in demand
- Potential conversion of a portion of the existing domestic well users to NUA water service, and potential future wholesale water sales to neighboring water providers
- Challenges with existing groundwater sources meeting more stringent Safe Drinking Water Act standards (e.g., more stringent arsenic standard promulgated in 2001, and potential new hexavalent chromium regulations in the coming years)

- Anticipated reductions in the estimated firm yield of Lake Thunderbird that will result in a reduction in the permitted amount of annual reservoir supply to Norman
- Potential climate change implications on supply and demand, ranging from projections that suggest that conditions could trend toward warmer and wetter, to hotter and drier for central Oklahoma and its upstream watersheds to the northwest
- Recent years' conservation successes and implementation of Norman's 2011 Water Conservation Plan, as well as implementation of water reuse for irrigation and conversion of certain groundwater wells to nonpotable service, all contributed to downward trends in per capita water use in Norman and in peak-day demand ratio (calculated as peak-day demand divided by annual average demand for a given calendar year)

The development of the 2060 SWSP drew on a range of sources, starting with the previously developed 2040 SWSP. Having already implemented many of the previous SWSP recommendations, NUA built on the 2040 SWSP. It also leveraged data and resources made available by the Oklahoma Water Resources Board's statewide water plan, the 2012 *Oklahoma Comprehensive Water Plan* (OCWP), providing insights into regional supplies and anticipated competition for supplies between multiple future users of each. The OCWP was also used in estimating passive and active conservation savings potential.

Development of the 2060 SWSP was guided by significant public input, and was actively solicited and brought forth at three levels:

1. Public meetings and Web-site materials (general public)
2. Ad Hoc Committee, appointed by City Council to represent key interests in a steering committee role for the SWSP planning team
3. City Council input and approvals.

The 2060 SWSP included supply option identification, screening, and packaging into supply "portfolios" for characterization and ranking. Supply options ranged from existing and new local sources, to additional water reuse for nonpotable demands, to augmenting Lake Thunderbird with highly treated water from Norman's Water Reclamation Facility as indirect potable reuse, to water conservation and new sources of supply, to new regional partnerships for regional supplies. With significant public input, the supply options were screened to those that passed fatal-flaw criteria and were the most cost-effective. Those options were then packaged into water supply portfolios for meeting the annual and peak-day demands through 2060, and evaluated relative to detailed criteria that were reviewed and modified through the public involvement process. The culmination of the 2060 SWSP is a phased, adaptive, water supply implementation plan.

REFERENCES

AWWA (American Water Works Association). 2013. Public Involvement and Customer Communication. www.awwa.org/about-us/policy-statements/policy-statement/articleid/209/public-involvement-and-customer-communication.aspx (accessed June 4, 2016).

AWWA (American Water Works Association). 2014a. Climate Change, approved Jan. 19, 2014. www.awwa.org/about-us/policy-statements/policy-statement/articleid/192/climate-change.aspx (accessed June 4, 2016).

AWWA (American Water Works Association). 2014b. Developing and Managing Water Resources, most recently revised and approved Jan. 19, 2014. www.awwa.org/about-us/policy-statements/policy-statement/articleid/194/developing-and-managing-water-resources.aspx (accessed June 4, 2016).

Field, P. 2010. *Facilitation…or Something More?* CBI Reports. Volume 13. Cambridge, Mass.: Consensus Building Institute. pp. 1–2.

IAP2 (International Association for Public Participation). 2012. Home page. www.iap2.org/ (accessed June 4, 2016).

NUA (Norman Utilities Authority). 2001. *2040 Strategic Water Supply Plan*. Norman, Okla.: NUA.

NUA (Norman Utilities Authority). 2014. *2060 Strategic Water Supply Plan*. Norman, Okla.: NUA. www.normanok.gov/sites/default/files/Utilities/images/2060%20SWSP%20-%20August%202014%20Final%20-%20Report%20Only.pdf (accessed Sept. 29, 2016).

OWRB (Oklahoma Water Resources Board). 2012. *Oklahoma Comprehensive Water Plan*. Oklahoma City, Okla.: OWRB.

Schwartz, R. 2002. *The Skilled Facilitator: A Comprehensive Resource for Consultants, Facilitators, Managers, Trainers, and Coaches*. San Francisco: Jossey-Bass. p. 5.

Smith, J., C. Howe, and J. Henderson. 2010. *Climate Change and Water: International Perspectives on Mitigation and Adaptation*. Denver, Colo.: AWWA and International Water Association.

USACE (US Army Corps of Engineers). 2010. *Responding to National Water Resources Challenges*. www.building-collaboration-for-water.org/Documents/nationalreport_final.pdf (accessed June 4, 2016).

USEPA (US Environmental Protection Agency). 2012a. Climate Ready Water Utilities. water.epa.gov/infrastructure/watersecurity/climate/ (accessed June 4, 2016).

USEPA (US Environmental Protection Agency). 2012b. Environmental Justice. https://www.epa.gov/sites/production/files/2015-03/documents/ips-consultation-guide_0.pdf (accessed June 4, 2016).

USGS (US Geological Survey). 2008. *Ground-Water Availability in the United States*. USGS Circular 1323. Denver, Colo.: USGS Information Services.

USGS (US Geological Survey). 2012. *National Water Information System*. waterdata.usgs.gov/nwis (accessed June 4, 2016).

WUCA (Water Utility Climate Alliance). 2010. Decision Support Planning Methods: Incorporating Climate Change Uncertainties Into Water Planning. www.wucaonline.org/assets/pdf/pubs_whitepaper_012110.pdf (accessed June 4, 2016).

Chapter 3

Legal Issues With Water Supplies

WATER RIGHTS IN GENERAL

Water is among the most fundamental concerns of any society. It is, therefore, unsurprising that the regulation of water and water rights has a long and tortuous history as governments decide whether and how to regulate this essential resource. In addition to natural concerns over water quantity and quality, policy judgments also play a role in this decision. Across the United States and throughout history, water law and policy have created a dynamic, continually evolving area of law.

Water rights are defined by statute, by regulation, and by common-law or other tradition. Current water rights and allocation methods reflect the differing conditions and historical influences that are present in different areas of the country. This chapter focuses on three distinct sources of water rights: common law rights, state rights, and federal statutory rights. The appropriation doctrine present in the western United States and the riparian system in the eastern United States illustrate how different approaches to water management are adapted to distinct environments and reflect historic influences.

From a scientific perspective, groundwater and surface water are not distinct resources but are interconnected pieces of the same hydrologic cycle. Traditionally, however, state water laws have addressed groundwater and surface water rights as though the two were separate sources of freshwater. Although this view has changed in some jurisdictions, the law in most states still fails to reflect the hydraulic connectivity of ground water and surface water. This disconnect creates conflicts among water rights holders and the regulators who are confronted with fairly allocating limited resources. It also creates the potential for over-allocation of water supplies in many jurisdictions, which is recognized as a current problem by the US Geological Survey (USGS).

Reflecting the range of influences that shape water allocation methods, water rights programs have long been the province of state—not federal—government. Some states delegate responsibility for water management to local governments, acknowledging that the nuances of water management may be best served at the local level. At each level of water management, a regulatory body may rely on common law doctrines or court decisions to resolve private disputes. Other state governments may develop sophisticated statutory and administrative arrangements to define individual water rights and to allocate limited water resources.

Ideally, water rights and management systems should be designed to avoid and resolve conflicts between competing uses of water as they occur. To accomplish this, water rights should be

- clearly defined and quantified,
- applied consistently to prospective and current water rights holders, and
- assigned to maximize the benefit of the resource.

The problem is how to define what "maximizes the benefit of the resource." What constitutes maximal benefit often depends on an individual's viewpoint. Regardless of the mechanism of determining maximal benefit, a water use allocation method that creates predictability allows water managers to rely on water rights when planning for future demands. By contrast, water rights systems without clearly defined allocation methods, predictability of use, or protection from use or waste by others tend to fail to meet societal demands.

Changing the structure of a state's water rights system can often be a long and complex process. However, when considering how to structure a water rights method, it is important to consider the underlying and sometimes competing interests of consumptive use, efficiency, conservation, historical allocation, and environmental needs. While no state's allocation system perfectly balances all these competing interests, water managers should be alert to opportunities to further align a state's water rights system with the ever-increasing demands for freshwater.

This chapter offers an overview of the existing water rights methods in the United States. It is by no means a detailed study but is designed to provide the reader with a basic background in the fundamental tenants of water law.

SURFACE WATER RIGHTS

Surface water rights in the United States were traditionally divided into two distinct systems: riparian rights in the East and the appropriation doctrine in the West. A more modern picture presents three systems of water rights: (1) riparian* rights, (2) regulated riparianism (which projects a system of permits and government regulation on top of the traditional court-made riparian doctrine), and (3) the appropriation doctrine (also called "prior appropriation"). A comparison of state surface water systems is presented in Table 3-1.

Riparian Rights

Overwhelmingly, riparian rights are used to allocate water in the eastern United States, roughly east of the Mississippi River. Riparian policies evolved naturally in an environment where water was generally plentiful and excessive government involvement was unwanted. The riparian doctrine, sometimes called the "rule of reasonable use," is patterned after water policy in England, where water resources look much like those of the eastern United States.

* *Ripa* means "riverbank" in Latin.

Table 3-1 Comparison of state surface water systems

State	Riparian	Regulated Riparian	Prior Appropriation
Alabama		X	
Alaska			X
Arizona			X
Arkansas		X	
California			X
Colorado			X
Connecticut		X	
Delaware		X	
Florida		X	
Georgia		X	
Hawaii		X	
Idaho			X
Illinois	X		
Indiana		X	
Iowa		X	
Kansas			X
Kentucky		X	
Louisiana	X		
Maine	X		
Maryland		X	
Massachusetts		X	
Michigan	X		
Minnesota		X	
Mississippi		X	
Missouri	X		
Montana			X
Nebraska			X
Nevada			X
New Hampshire	X		
New Jersey		X	
New Mexico			X
New York		X	
North Carolina		X	
North Dakota			X
Ohio	X		
Oklahoma			X
Oregon			X
Pennsylvania	X		
Rhode Island	X		
South Carolina	X		
South Dakota	X		
Tennessee	X		
Texas			X
Utah			X
Vermont	X		
Virginia		X	
Washington			X
West Virginia	X		
Wisconsin		X	
Wyoming			X

Under the riparian doctrine, the right to use water from a lake or stream belongs to whoever owns the land abutting the water source. Such a landowner is a riparian owner. Every riparian owner is entitled to use water from the stream. This right is defined as the right to enjoy the advantage of a reasonable use of the stream as it flows through the landowner's property. This right, however, belongs equally to all riparian owners.

Two rules have governed how much water a riparian owner may use. The older rule held that a landowner must leave the natural flow of the river unchanged. Each downstream riparian was entitled to the water in its natural condition, without others' altering the rate of flow or the quantity (or quality) of the water. The modern rule of reasonable use allows each riparian owner to use the water, regardless of the natural flow, so long as the riparian's use does not cause an unreasonable injury to any other riparian.

Regulated Riparian Rights

Over time, population growth and development in the East have increased the problems of water distribution historically addressed by riparian rights. This phenomenon and an increased faith in government regulation have caused most states to overlay the traditional riparian system with new administrative methods, such as permit systems, to regulate water use. These "regulated riparian" systems combine aspects of the traditional riparian doctrine (e.g., reasonable use) with governmental regulation. Currently, more than half of the states east of the Mississippi River employ some version of regulated riparianism for managing surface water.

The most important feature of regulated riparianism is that users must have a permit to divert and use surface water. Although the standard for granting the permits is typically similar to the standard for reasonable use, it is not identical and may vary by state. Commonly, the water permitting process is delegated to local or regional governmental entities within a state.

Regulated riparian systems are typically created by statute. These statutes usually contain a statement of purpose that lists goals such as the conservation of water, promotion of beneficial use, or protection of the public welfare. Because there is no uniform, regulated riparian code, however, the precise application of regulated riparianism varies widely from state to state. For instance, many states exempt domestic and agricultural uses from the permit requirement, while others simply assign domestic use the highest priority. Some states contain no domestic preference at all.

As water conflicts grow in frequency and severity across the country, many states are turning away from the common-law doctrines of absolute ownership, reasonable use, and correlative rights. A concept that sometimes appears in groundwater regulation, the rule of correlative rights holds that the right to make overlying use of water is not absolute but is relative to the rights of other overlying users. By overlaying modern concepts of administrative permitting on a background of common law doctrine, regulated riparian states are increasingly able to tailor water rights systems to the specific, regional water needs of the state. Given the water conflicts already present in much of the south and west of the country, it seems likely that regulated riparianism will continue to gain prominence as a flexible, effective water management method.

Prior Appropriation

Although the verdant eastern states benefit from increased rain and the riparian system, water rights in the arid western United States require a different approach. Historically, trappers, miners, and settlers who migrated west encountered a hostile, dry environment. Early explorers referred to the Great Plains as the Great American Desert, and not all believed that it could be settled. It was obvious that most of the land would require irrigation. Limiting the use of streams to only adjoining landowners was not practical; such an action would drastically curtail the development of the new lands because "nonriparian" land would be practically useless.

The early miners are credited with finding a solution to the problem. By custom, they all accepted the fact that the first miner who used water from a stream to work his claim was protected against latecomers. Soon this custom expanded to include the use of water for all purposes, not just for mining. Finally, as the land was organized into territories and then into states, the custom became law through express recognition by court decisions, constitutional provisions, and state statutes.

What emerged was a flexible and useful concept—"prior appropriation"—designed for western conditions and tempered by the western experience. The federal government promoted prior appropriation by enacting legislation that encouraged settlement of the West in conformity with each state's water rights laws.

The result of this flexible approach was that no western state completely adopted the riparian doctrine. About half, bowing to logic and precedent, retained or modified some riparian features. Other states repudiated riparianism altogether. Those states that originally retained some riparian features have found that, over time, riparianism's vestiges prohibit the full development of limited water resources. These states continually modify their water law to reduce the effect of riparian principles. Although prior appropriation remains critically important to the western United States, the doctrine can result in conflicts and costs for public water utilities that may need to acquire water rights from current owners. All 17 contiguous western states and Alaska embrace the main principles of the appropriation doctrine. Hence, the appropriation doctrine is firmly established as the principal rule of water resources in the West.

The appropriation doctrine envelops several interrelated concepts:

- Water that is available for appropriation
- A permittee that has (or will) put appropriated water to a beneficial use and not waste it
- The physical act of impounding or diverting water

In times of shortage in a prior appropriation system, a senior water rights appropriator has priority to withdraw water before any other junior appropriators. In other words, the oldest rights prevail. The prior appropriation system grants a holder a right of use or *usufructuary right*. While title to waters in a flowing stream remains in the name of the public or the state, those waters may be used for the benefit of the people.

The concept that public waters must be used for a useful or beneficial purpose lies at the heart of western water law. Thus, if an appropriator has no immediate use for the full appropriation, the law requires that only the amount presently needed be diverted, allowing excess water to remain in the stream. Similarly, once water has served the appropriator's beneficial use, any waste or return flow must be allowed to return to the stream. This requirement protects not only the public but other appropriators as well.

A water right is acquired by taking waters and applying them to a beneficial use. In some states, an appropriator need not obtain permission to put water to beneficial use. However, in most states, water may not be taken or used until the user has obtained a permit from a water administration official, usually the state engineer or environmental agency.

As previously discussed, senior water users have priority over later appropriators during times of water shortage. Obviously, each appropriation on a stream affects all others on the same stream. Consequently, states have developed various systems to determine the relative rights and priorities of water users. These duties commonly fall to an administrative agency, whose actions are subject to court scrutiny, or are conducted through special court proceedings, resulting in court decrees that establish the relative rights and priorities of all water users from the same supply.

According to the USGS, there remains a concern that the western states have over-allocated their water resources. See, for example, USGS State Water Resources Research Institute Program's abstract in *Benefits and Costs of Water Reuse in Texas: Abstract* (USGS 2011). Over-allocation creates the potential for significant future conflicts among competing water uses, including environmental flows.

GROUNDWATER RIGHTS

Water rights systems for groundwater can also be categorized into three major common-law doctrines: absolute ownership, reasonable use, and correlative rights. Although few states use one of these doctrines without some modification, each doctrine can be found in the United States. In some states, more than one doctrine may be in use.

Table 3-2 reflects the major common-law doctrines that form the basis of the groundwater method in place in each state. This table also indicates where a prior appropriation system is in place for state groundwater. As discussed below, states have adopted numerous variations on each of the three common-law doctrines, creating hybrid systems that do not always fit neatly into one of the three major doctrines. These subtleties are *not* reflected in Table 3-2.

The common law rule of absolute ownership (also called the "English Rule") is the oldest of the doctrines applied to groundwater and stems from the principle that land ownership encompasses everything beneath the land to the center of the earth. Of course, the migratory nature of groundwater distinguishes it from most substances found beneath the earth and makes it impossible to apply the concept of ownership in a literal sense. Practically, the rule of absolute ownership is a rule of capture; a landowner may use all the groundwater that can be captured from beneath that owner's land. Thus, when a conflict develops, the person with the deepest well and the biggest pump prevails.

As with the rule of absolute ownership, the rule of reasonable use (also called "the American Rule") holds that groundwater rights are an incident of land ownership. A landowner may generally withdraw and use groundwater but is limited to withdrawing groundwater for beneficial use. Under this doctrine, a groundwater user whose use interferes with that of neighboring landowners may continue to withdraw groundwater only if his use is reasonable. Reasonable use precludes wasteful or nonbeneficial uses of groundwater. Furthermore, a groundwater user can incur liability from harm caused to others by an unreasonable use of groundwater (i.e., subsidence). Thus, the doctrine of reasonable use confers only a qualified right, rather than an absolute right, to use groundwater.

The rule of correlative rights holds that the right to make an overlying use of water is not absolute but is relative to the rights of other overlying users. The rule is used primarily when the groundwater supply is insufficient to satisfy the needs of all overlying users and sharing is required. In its strict application, correlative rights mandates proportional sharing of groundwater on the sole basis of overlying acreage. However, few states apply a pure version of correlative rights. Most states choose, instead, to temper the doctrine with considerations of reasonableness or net public benefit.

In practice, most states employ a hybrid system of these groundwater management systems rather than adhering strictly to one of the doctrines discussed above. These hybrid methods of groundwater management are sometimes referred to as "regulated riparian" systems and bear many similarities to the surface water management system by the same name. For instance, many states purporting to follow the common law doctrine of reasonable use or correlative rights also require permits for groundwater withdrawals. Similarly, absolute ownership states may also maintain local management districts that

Table 3-2 Major common-law doctrines in 50 states

State	Prior Appropriation	Reasonable Use	Correlative Rights	Absolute Ownership
Alabama		X		
Alaska	X			
Arizona		X		
Arkansas			X	
California			X	
Colorado	X			
Connecticut				X
Delaware			X	
Florida		X		
Georgia		X		
Hawaii			X	
Idaho	X			
Illinois		X		
Indiana				X
Iowa		X		
Kansas	X			
Kentucky		X		
Louisiana				X
Maine				X
Maryland		X		
Massachusetts				X
Michigan		X		
Minnesota			X	
Mississippi				X
Missouri			X	
Montana	X			
Nebraska			X	
Nevada	X			
New Hampshire		X		
New Jersey			X	
New Mexico	X			
New York		X		
North Carolina		X		
North Dakota	X			
Ohio		X		
Oklahoma			X	
Oregon	X			
Pennsylvania		X		
Rhode Island				X
South Carolina		X		
South Dakota	X			
Tennessee		X		
Texas				X
Utah	X			
Vermont		X		
Virginia		X		
Washington	X			
West Virginia		X		
Wisconsin		X		
Wyoming	X			

issue permits for withdrawals that occur within a certain geographical or hydrological area. For example, Texas has groundwater management districts regulating production in more than two thirds of its counties. Finally, some states incorporate aspects of prior appropriation into the groundwater method, such that the person who first withdraws groundwater retains the right to do so against future users. This version of prior appropriation is essentially identical to the prior appropriation system for surface water.

RECLAIMED WATER RIGHTS

Water reuse has recently attracted attention as a source of water, particularly in the western part of the United States as droughts persist. Once considered a waste or by-product, reclaimed water has now become a consistent source of water supply for many water users.

The rights to reclaimed water vary across the country. The rights also vary depending on the underlying source of supply. Most states treat reclaimed water as being surplus to the diversion and use of state water, and once returned to a watercourse, the rights have essentially been abandoned unless there is a claim to the return flows. Even if there is a claim to return flows, there must be some form of a "no injury" analysis performed on surface water rights and the environment to ensure there is no negative effect by allowing for the subsequent rediversion and reuse of reclaimed water. In Texas, for example, the protocol involves securing what is considered a "bed and banks" authorization to secure the rights to convey, redivert, and reuse reclaimed water that is discharged. In addition, just as in the context of the differing legal treatment of groundwater and surface water, the Texas permitting system treats the analysis of the no-injury assessment differently depending on the underlying source of supply.

The rights to reclaimed water that is not discharged, relinquished, or otherwise conveyed to a third party generally reside with the holder of the underlying supply absent some contractual provision to the contrary. For example, many wastewater treatment contracts include a provision related to the transfer of liability between a retail system (e.g., a city) and its wholesale provider (e.g., a regional provider) that often contains language related to transfer of liability, including title. In those contractual instances, it is clear that the conveyance of title to the raw wastewater has occurred, and absent any reservation of rights to the reclaimed water, title to the reclaimed water continues to rest with the wholesale provider post-treatment.

FEDERAL AND TRIBAL WATER RIGHTS

Federal water rights consist of federal reserved rights, water rights for water not reserved, rights of navigable water, and federal contract rights.

Reserved Rights

The federal reserved water rights doctrine, frequently referred to as the "Winters doctrine" after *Winters v. United States*, states that when the United States sets aside or reserves a part of its lands for particular uses or purposes, it reserves—by implication—the right to enough of the unappropriated waters on or adjacent to the lands to meet its uses and purposes. This implied reservation usually takes priority at the time the lands are reserved.

Although the rights reserved by the United States are junior to the rights of prior appropriators, they differ from appropriated rights in several ways: (1) the waters need not be put to use promptly or within any particular period of time; (2) although the rights are ultimately subject to measurement, they need not be quantified at the time the reservation is established or within any particular period of time; (3) the rights are not subject to loss by nonuse or abandonment; and (4) the water may be used for purposes other than those for which the land was reserved, as long as the alternative uses do not increase consumption.

Although beneficial use, in relation to the purposes for which the reservation was created, ultimately determines the measure of reserved rights, a reserved water right is not limited to the purpose for which it was created. Rather, the water may be put to other beneficial uses as long as those uses do not increase consumption. In short, particular beneficial use is the measure, but not the limit, of a reserved water right.

Water Rights for Federal Lands Not Reserved

Not all federal lands are reserved. The Bureau of Land Management administers millions of acres of unreserved public lands under congressionally authorized programs. These programs may require water.

Does the government have water rights under federal law that it can use to meet congressionally authorized needs of unreserved lands? The question has raised a great deal of debate. Central to the debate is the issue of state control over appropriations. Some commentators contend that the federal government must comply with state law in acquiring water rights for use on unreserved lands; others contend the federal government can establish the rights based on congressional mandate. Between 1979 and 1981, three attorneys offered three opinions on the scope—and, indeed, the existence—of these rights. The US Supreme Court, in a decision that predated the attorneys' opinions, directed the federal government to comply with state law, but only as far as the state law is consistent with Congress's objectives. At least one earlier court case supports the government's right to use water on its unreserved lands.

Federal Rights on Navigable Waters

Under the commerce clause of the US Constitution, the United States retains control over all navigable waters, in the interest of navigation. In this context, the United States has a dominant servitude over such waters and the beds and banks of lakes and streams that contain them. All proprietary rights of others related to such water and the land beneath it are subject to this dominant servitude or easement for navigation and may be impaired or extinguished by the United States without compensation.

Contractual Water Rights

Federal contract rights usually relate to the US Bureau of Reclamation (USBR) and the US Army Corps of Engineers (Corps). Under the Reclamation Act of 1902, the Secretary of the Interior, acting through the USBR, is authorized to construct, operate, and maintain water storage and distribution facilities in the 17 western states. The USBR is also authorized to make water from such projects available to users for irrigation and other purposes. Section 8 of the act directs the Secretary of the Interior to act "in conformity with [state] laws relating to the control, appropriation, use, or distribution of water used in irrigation," 43 USC § 383 (2012). The act also states that "the right to the use of water acquired under the provisions of this act shall be appurtenant to the land irrigated, and beneficial use shall be the basis, the measure, and the limit of the right," 43 USC § 372 (2012).

To acquire water rights, the USBR usually files an application to appropriate a specific amount of unappropriated water in a river, and/or acquires vested water rights by purchase or condemnation. After constructing the dam and storage reservoir and, if necessary, the distribution works to deliver project water to individual users, the USBR executes after-delivery contracts with irrigation districts or other similar entities, according to the terms of the Reclamation Project Act of 1939.

The USBR may also enter into contracts to provide water for municipal and industrial uses. These contracts make project water available in return for the users agreeing to repay a portion of the construction, operation, and maintenance costs over a specified period.

The Corps' historic mission to maintain and improve navigability on US harbors and navigable rivers was gradually broadened during the early 1900s to encompass flood control. Beginning in the 1930s, the Corps was authorized to construct dams to control floods and regulate streamflows to maintain navigability. The Secretary of the Army has no authority to provide storage and delivery of water for irrigation, but Congress authorized the provision of irrigation water in a number of Corps projects under the supervision of the Secretary of the Interior pursuant to the reclamation laws.

Interstate Compacts

Interstate compacts, like contracts, govern the apportionment of water resources between or among states. An interstate compact must be negotiated, ratified, and signed by the states involved. Once approved, the compact becomes a binding federal statute. Aside from these universal features, interstate compacts vary widely. For instance, most (but not all) interstate compacts create independent governmental entities called "commissions," consisting of representatives from each state, to make rules, develop policies, and oversee the administration of water rights under the compact. Some compacts address issues of water pollution, flood control, or resource management, while other compacts are limited to questions of water allocation. Some compacts grant enforcement authority to the compact commission, while others do not. A few compacts, including the Rio Grande Compact, even affect groundwater use. The Colorado River Compact affects the states of Colorado, Utah, Nevada, Arizona, and California. This compact is responsible for a significant portion of the water delivered to homes and utilities in Nevada, Central Arizona, and Southern California.

Unlike typical contracts, interstate compacts governing water resources must receive Congressional approval to satisfy the compact clause of the Constitution. There are currently more than 20 approved interstate compacts that address water allocation between states. Most of these compacts are located in the arid regions of the country.

Additionally, four compacts, designated "federal interstate compacts," include the signature of the United States as party to the compact. These federal interstate compacts include the Delaware River Basin and the Susquehanna River Basin Compacts. The federal interstate compacts tend to involve significantly more commission oversight over individual diversions and activities that may affect the basin.

Tribal Rights

Tribal rights are created by statute, executive order, or court doctrine. The rights generally can be divided into two types of water rights: aboriginal (Winans) rights and federal (Winters) rights. An aboriginal water right has seniority and, therefore, priority over other appropriators in the basin, but the right must be proven by showing actual use. The use may be nonconsumptive, such as hunting or fishing, but an aboriginal water right cannot be transferred to another person or a new use without losing its priority date from "time immemorial."

By contrast, federal rights are considered incident to the federal government's creation of a Native American reservation. Because reservations were originally designed to transform native populations from nomadic hunters into farmers, a quantity of water sufficient for irrigation was required for each reservation. This quantity is determined by assessing "practicably irrigable acreage" on the reservation. However, in recent years, the "practicably irrigable acreage" standard has received significant criticism and been the subject of intense federal litigation.

What rights included in a tribal water right can vary widely from one state or court case to another. In some cases, courts have found that tribes are entitled to a certain quality of water as well as a designated quantity. Some courts have allowed tribes to claim groundwater rights. Other courts have allowed tribes to enjoin others from withdrawing groundwater in a way that interferes with tribal surface water rights. A water manager whose work may affect tribal water rights should be aware that this area of the law is fraught with conflicting policies.

PUBLIC TRUST DOCTRINE

The Public Trust Doctrine is a legal framework for resource planning and management that has been used to clarify water rights. Its origins are historical, and the doctrine is a currently evolving concept relating to the ownership, protection, and use of essential natural and cultural resources. The doctrine is receiving increased attention in the United States because of the growing awareness of the duty of care owed the environment. The Public Trust Doctrine is a common law doctrine that emphasizes the legal right of the public to use certain lands and waters. The right may be concurrent with private ownership. Under the Public Trust Doctrine, legal interest of the public is not absolute; it is determined by a balancing of interests.

History

The Public Trust Doctrine has its roots in the Institutes of Justinian, the body of Roman civil law assembled by the Roman Emperor Justinian's top legal scholars in 530 CE. One of these laws stated that running water, the sea, and consequently, the shores of the sea were common to all of humanity. Therefore, no one was forbidden from approaching the seashores. England, in adopting much of the Roman law, recognized waters and shores as public in nature. As commerce became more important, so did the public's interest in the shores. Eventually, the shores came to be recognized as property owned by the king in trust for the public.

In America, as the colonies began to acquire charters from the kings of England, the law of public shore lands came to America. The waterways were so vital for commerce and sustenance that the original 13 states kept the shore lands under their control when they granted the Northwest Territory to the federal government. As the rest of the states joined the Union, they did so on an equal footing with the original 13 and kept the shores under their control. This land was deemed to be owned by the states in trust for the public, hence the name Public Trust Doctrine.

Application to Water Rights

Under the Public Trust Doctrine, title to water resources or the corpus is held by the state, as sovereign, in trust for the people. The purpose of the trust is to preserve resources in a manner that makes them available to the public for certain public uses.

Until recent decades, the predominant corpuses recognized as subject to the Public Trust Doctrine were tidal and navigable waters. American cases have held that title to lands underlying tidal and/or navigable waters is held by the state in its sovereign capacity as trustee for the benefit of the citizens of the state who have the right to use the waters and adjacent land for navigation and to "fish, hunt, or bathe." In other words, tidewaters to their farthest reach, tidelands, navigable waters, and permanently submerged lands, including those extending lakeward or seaward to the limit of state ownership, are subject to the Public Trust Doctrine.

State Regulation

The Public Trust Doctrine originated as an instrument of federal common law used to ensure protection of the public's interest in navigation, fishing, and recreation (except in Maine and Massachusetts, where use is limited to fishing, fowling, and navigation). In addition, many state constitutions (e.g., Hawaii) incorporate the doctrine, which declares, "All public natural resources are held in trust by the State for the benefit of the People." This allows the legislatures to regulate and control submerged lands underlying tidal and navigable waters through the enactment of statutes and ordinances. These laws then enable the states to more easily resolve disputes within their own courts.

The Public Trust Doctrine has also been used in the area of water rights by individual states, particularly California. In 1983, the California Supreme Court used the Public Trust Doctrine to address the diminished water levels in Mono Lake in *National Audubon Society v. Superior Court*. In this case, the Court determined that the state of California had an affirmative duty to consider the public-trust value of Mono Lake when permitting initial allocations of water and when reviewing past allocations. The Court then required examination of the diversions of tributary flows to Mono Lake by the City of Los Angeles, and the state ultimately reduced the amount of diversion permissible under California law. Since this decision, Mono Lake levels have increased significantly.

Some states, in addition to California, have also applied and even expanded the Public Trust Doctrine in the context of water law. For instance, Hawaii expanded the Public Trust Doctrine to include not only surface water rights but also groundwater rights. Similarly, the state of Vermont codified the Public Trust Doctrine so that the doctrine explicitly applies to groundwater. Even in eastern states, such as Connecticut, groundwater withdrawals are limited by regulation to prevent effects to streamflow during critical periods. Other states, however, have limited the scope of the Public Trust Doctrine. Although Idaho's judiciary applied the doctrine to state water rights, the legislature overruled that decision and embodied the state's rejection of the Public Trust Doctrine, as applied to water rights, in statute.

The Public Trust Doctrine has been a primary tool of instream flow interests attempting to restore over-appropriated rivers and streams. Although the Mono Lake water levels were restored because of a court's application of the Public Trust Doctrine, the doctrine has been met with mixed success in other situations. This is partially because the power of the Public Trust Doctrine depends heavily on language contained in the relevant state's constitution. In some instances, even if the Public Trust Doctrine is not applied, courts may retroactively apply a "public interest" test to determine whether water diversions are permissible, thereby accomplishing the same goal as the Public Trust Doctrine.

WATER QUALITY EFFECTS

Traditionally, water rights and water quality regulations have been viewed as separate and distinct creatures of law. Water rights are founded on the principles of preventing waste and requiring beneficial use of the resource. Regulations affecting water quality are based on the principles of limiting or eliminating effects, and protecting existing beneficial uses of the resource. They also address remediation of contamination so that beneficial uses can eventually be attained (e.g., total maximum daily load requirements). Although their paths have different origins, water rights and water quality regulations can, and often do, converge.

Early on, stream quality was not affected by the use of water. Initial demands for water were simply not that great. As the population has continued to grow, however, and more demands have been placed on existing resources, water quality effects must be considered. This is no more evident than in the debate over environmental flows. Over the last two decades, many western states have grappled with the issue of granting water rights for consumptive needs while still protecting environmental flows. The delicate balance between human and aquatic needs is something that many states are still struggling to achieve. While the "use" of water generally consists of diversion and consumption, some consideration must be given to leaving a portion of water in its natural flowing state. Recently, regulations have begun to reflect the economic value in maintaining sufficient streamflow to support the local flora and fauna.

Select pieces of federal legislation seek to balance water use and water quality regulations. The Clean Water Act (CWA) and the Endangered Species Act (ESA) have both played a role in the use of water resources. The CWA has affected water rights in the environmental flows arena, whereas the ESA may curtail the use of water to protect endangered species. Both the CWA and the ESA have sought to balance environmental needs with human needs, but there have been prolonged legal battles to define the exact meaning of these laws.

Instream Uses and Environmental Flows

Water suppliers across the arid West are well aware of the term "environmental flows." Generally considered the amount of base flow needed to protect riverine and estuary health, environmental flows are essential to tourism, fishing, and ecology industries across the country. The debate over environmental flows has not so much focused on whether some amount of flow is necessary to protect aquatic health, but more about who is authorized to hold rights for environmental flows, and how much flow is necessary for each stream segment or estuary.

Of the western states that follow the appropriation doctrine, the majority address environmental flows by reserving such water (i.e., reservation) from appropriation. Although some states allow one agency to assign an amount of water to another state agency, this action is still consistent with reservation. The intent of reservation is for the state to withhold some amount of water, which would otherwise be available for appropriation, and to reserve that water in the public trust for the benefit of the environment. There are some states that allow individuals, including nonprofit organizations, to apply for and receive appropriative rights for environmental flows, but there is great debate over whether such action actually constitutes an appropriation for physical storage or a diversion of water for beneficial use.

Clean Water Act

By establishing water quality standards, the Clean Water Act affects water rights. A water quality standard sets a goal for water quality by designating permissible uses of a water body and setting numeric criteria that will maintain that use. Whenever attainable, the CWA water quality standards aim to provide for both the protection and propagation of fish and wildlife and also for recreation. Tailored water quality standards form the foundation of the CWA's pollution controls. For example, the level of treatment technology required for wastewater treatment plants varies in accordance with the water quality standards applicable to different stream segments. Total maximum daily load criteria can be used to ensure that pollution prevention meets the needs of specific stream segments.

Water quality standards may affect the diversion and use of water resources. For instance, stream segments with low dissolved oxygen or elevated temperatures may need to maintain higher base flows to maintain aquatic health. One tool that can be used to maintain base flow is to limit diversions during low-flow events. Many western states employ some form of streamflow restrictions, or special conditions, when granting appropriative rights to maintain water quality standards.

In addition to setting water quality standards, the CWA also contains a permitting regime. The National Pollutant Discharge Elimination System (NPDES) requires authorization before any person may discharge pollution into a water body from a point source. "Point sources" include wastewater treatment plants, stormwater systems, industrial facilities, and agricultural sources. Because many water supply projects contemplate moving water from one of these point sources to another (e.g., moving reclaimed water from a wastewater treatment plant to an industrial facility), water supply projects may require authorization under the NPDES permitting regime.

Endangered Species Act

Congress enacted the Endangered Species Act in 1973 to counter the population decline and potential extinction of numerous animal and plant species. Over the last two decades, the ESA has generated conflict over the use of private property inhabited by endangered species. The ESA reaches private control of land and resources through its general prohibition against the taking of endangered species. Under the ESA, "the term 'take' means to harass, harm, pursue, hunt, shoot, wound, kill, trap, capture, or collect, or to attempt to engage in any such conduct," 16 USC § 1532(19) (2013). "Harm" is the broadest and least-precise term in this definition.

In the water rights context, the ESA has affected the use of both surface and groundwater. In Texas, the ESA has been used to impose pumping limits in the Edwards Aquifer to protect the fountain darter. In New Mexico, the ESA has been used to limit diversions of surface water to protect the silvery minnow in the Middle Rio Grande. As part of the habitat for endangered species, the ESA recognizes the flow of water, either underground or on the surface, as being imperative to the survival of aquatic species. Recognizing that certain water supply projects may reduce the flow of water, the ESA has become an effective tool for restricting or limiting the diversion and use of water resources. It also requires consultation for any project that needs to obtain a federal wetlands permit under the CWA.

REFERENCE

USGS (US Geological Survey). 2011. Abstract. In *Benefits and Costs of Water Reuse in Texas*. USGS State Water Resources Research Institute Program. water.usgs.gov/wrri/2011grants/2011TX396B.html (accessed August 2016).

Chapter 4

Water Resources Alternatives

Of the total water found on the earth, 97.3 percent is saltwater in the oceans. Of the remaining water, more than two-thirds exists as ice in the polar caps. The rest, or 0.61 percent of all water, is freshwater in lakes, rivers, streams, and groundwater. That means that while there is a lot of water out there, getting to a sustainable supply may be of issue (AWWA 2014a, 2014b). This chapter identifies alternatives for developing, expanding, or maximizing a water resource portfolio. The water resources listed below are identified and described in this chapter; an evaluation of each resource is provided in chapter 8:

- Surface water
- Groundwater
- Reclaimed water
- Desalination
- Demand management
- Water marketing and transfers
- Nontraditional sources

SURFACE WATER

In most regions, surface water sources are viable candidates for meeting water supply needs. Generally, the four alternative surface water sources are as follows:

1. Direct river withdrawals
2. Onstream reservoirs
3. Pumped reservoirs
4. Stormwater

Direct River Withdrawals

This option requires either one of the following:

- A locally available source with sufficient base flow to allow dependable direct diversion throughout a wide variety of flow conditions (without significant diversion storage for release during low-flow periods: see onstream and pumped reservoirs sections below for these instances); or
- Availability of alternative sources that can independently be relied on to meet water supply requirements during periods when the river diversion source may be unavailable (i.e., conjunctive use of groundwater as discussed later in this chapter).

Unless a river source with a very large drainage area is available, it is likely that drought period flows will create less than minimum instream flow targets, requiring storage either in a reservoir or aquifer to provide source dependability. If storage is required, the source would be defined either as a pumped-diversion reservoir, aquifer recharge, or an aquifer storage and recovery source. Each is discussed later in this chapter.

Onstream Reservoirs

Onstream reservoirs (such as natural and human-made lakes) rely on usable reservoir storage capacity and natural basin runoff for meeting water supply needs. During periods of low-flow reservoir recharge, water supply diversion requirements are met or supplemented through reservoir drawdown. During periods of plentiful runoff into the reservoir, water supply needs can be met by streamflow with storage recharge provided by excess runoff. These facilities offer the benefit of low operating cost, because reservoir recharge is derived from natural runoff.

When located in upland areas, pumped transmission to treatment facilities can sometimes be minimized or eliminated with delivery by gravity pipe-flow or by stream channel releases with downstream diversion at or near the water plant site. Many times, main-stem (onstream) reservoirs will have a greater level of overall environmental impact than reservoirs on smaller tributary streams, because of typically increased wetland and habitat disturbance, impediments to fisheries resources, disruption to free-flowing streams (stream bifurcation), and other factors discussed in chapter 8, Water Supply Assessments.

Pumped Reservoirs

If significant pumped diversions are provided to a reservoir as a reserve for meeting demands during droughts or other periodic service interruptions, the project alternative is considered a pumped-diversion reservoir project. These facilities are also known in some regions as skimmer or *side-stream* reservoirs. For this project type, much of the storage recharge is provided by diversions from outside the reservoir's natural watershed, supplemented with natural basin runoff. The pumped-diversion reservoir option has many times been found to best balance environmental concerns with water supply development needs.

To better protect aquatic life and accommodate downstream consumptive uses, minimum instream flow requirements have been steadily increasing. These restrictions can significantly reduce both the viability and available yield of direct river withdrawals, onstream reservoirs, and pumped-diversion reservoirs. Pumped-diversion reservoirs are somewhat more capital intensive because of the need for pumped-diversion facilities (intake, pumps, and additional pipeline construction), but they are many times more responsive to overall environmental concerns for a variety of project settings. Additionally, many streams of sufficient drainage area for water supply purposes lack the appropriate topography and geology for developing a cost-effective dam and reservoir site for drought storage. Onstream reservoirs are also more likely to require a large and expensive spillway, and the valley areas of larger streams are more likely to have infrastructure development that would be affected by a reservoir. Where better settings for a dam and reservoir are found on small tributary basins, the source of supply and the storage facility can be considered separately to minimize overall development effects.

Stormwater

Stormwater capture is very similar to river withdrawals with the exception that there is typically no base flow for continuous capture. In most cases, diversion and storage is necessary to maximize capture from seasonal storms that can vary widely in duration and intensity. In California, rubber dams are frequently installed to divert storm flows into off-channel recharge basins (similar to onstream or pumped reservoirs) but can be lowered during periods of high flows to prevent upstream surcharge and possible overflow of storm channels.

GROUNDWATER

Groundwater is discussed herein as it applies to a water resources alternative; for more detailed information about groundwater, refer to AWWA Manual M21, *Groundwater* (AWWA 2014a).

Reilly et al. (2009) estimated that the pumpage of fresh groundwater in the United States is approximately 83 billion gallons per day (Hutson et al. 2004). Utilities are competing with agriculture and power, which use about 80 percent of all water withdrawals in the united States. Utilities are 12.5 percent of the total (Lisk, Greenberg, and Bloetscher 2012). More than half of the people served with water supplies in the United States and Canada obtain their water supplies from groundwater. Nearly 80 percent of all utilities, and most of the smaller utilities, derive their source water from groundwater. But groundwater is not visible from the surface, and the understanding of its behavior and occurrence among the public is limited. Reilly et al. (2009) noted the following:

1. *In contrast to rivers and lakes, ground-water systems are hidden from direct observation and measurement,*
2. *The sources of water to ground-water systems and the time required for the effects of withdrawals to propagate through the system and be observed are different for each system,*
3. *The amount of detail (spatial scale) needed to describe the resource depends on the objectives and purpose of the desired information,*
4. *The amount of change in ground-water levels that is important is different for different ground-water systems,*
5. *Not all water pumped is consumed and much of the water pumped is redistributed and changes the ground-water flow system, and*
6. *The chemical quality of the water is important in determining its suitability (and thus its availability) for various uses.*

Groundwater withdrawals can and usually do affect the amount (and quality) of surface water. Reilly et al. (2009) suggested that water use is about 8 percent of the estimated 1 trillion gallons per day of natural recharge to the nation's groundwater systems (Nace 1960). Reilly et al. (2009) found that groundwater supplies, however, are over-pumped in many areas and that the long-term results could be catastrophic, affecting economic viability of communities and potentially disrupting lives and ecological viability.

Groundwater Development

The development of groundwater is more complicated than that of surface water. Groundwater is a resource that

- is hidden from view;
- is not amenable to delineation of basin boundaries from a topographic map;
- is usually in a state of dynamic equilibrium;
- does not necessarily follow surface watershed divides;
- may comprise several aquifers in different geologic layers that have different areal extents, different flow directions, and different recharge or discharge areas;
- may have significantly different water quality in different vertical aquifers in the same topographic location;
- may have different hydraulic potential in different aquifers (such as confined and unconfined) in the same location; and
- is affected by the quality of sediment and rock materials through which it passes.

As a result, groundwater source identification, development, and continued management require consideration of numerous environmental factors.

The primary options for groundwater sources of supply include the following:

- Groundwater wells for direct withdrawal, or groundwater extraction wells, are holes drilled into the ground to a formation that will permit water to be extracted from it. These water-bearing formations are called aquifers. In most aquifers, extraction of water can only be accomplished through the use of pumps, because the groundwater surface is below the ground level. Artesian aquifers are aquifers that are under pressure. Artesian aquifers will flow to the surface, which may limit the amount of pumping needed.
- A collector well or subsurface drain (or horizontal well) intercepts interflow in permeable materials or infiltrating surface water and discharges into a sump whose bottom is below the invert of the gallery screen and casing.
- Infiltration galleries are often exposed to the surface but intercept groundwater.

Groundwater development often begins when an individual digs a hole in the ground, pumps the water out of the ground, and uses it for domestic purposes or for irrigating crop fields. As the population grows and additional holes are dug, the need for organized groundwater development arises in the face of reduced yield at the wells and poorer water quality.

Groundwater Extraction

A groundwater well is a hydraulic structure that allows water to be cost-effectively withdrawn from water-bearing formations (aquifers). The utility of a water well is not limited to water supply. Wells are also used to monitor water levels and water quality, drain agricultural lands, prevent saltwater intrusions, recharge aquifers, discharge wastes, and relieve pressures under dams and levees. Wells are also used for many other purposes including oil and gas exploration, hydraulic fracturing, hazardous waste disposal, and injection of a variety of other waste products.

Three steps are used in identifying potential groundwater sources for a public water supply. The first is identifying regions having low-pollution potential, high recharge capability, and a favorable location to the utility and its customers. The second step is performing field investigations to confirm site-specific characteristics of the aquifer system to determine how much water is potentially available. Care should be taken with confined aquifers where water withdrawal capability is consistently overestimated. The third is dealing with land-use and wellhead protection issues.

Evaluating regional groundwater conditions and the potential for resource development should be based on the following factors (AWWA 2014a):

- Quantity and quality of water required
- Availability of water resources within the vicinity of the regional demand, including the sustainable yield of the potential source aquifers
- Cost to develop the water supply for the region
- Nature and density of existing or likely future pollution sources within the recharge area of an aquifer, and the effectiveness of regulatory controls on these sources
- Amount of previous groundwater development and availability of hydrogeologic or groundwater pollution investigations
- Long-term land uses affecting quality and quantity of groundwater recharge
- Conveyance costs and treatment required for specific development projects

This information is seldom available to a planner. Often what is available at best, in addition to the regulatory and administrative standards, are desired yield, location of use, and approximate location of a well. This available information is supplemented by hydrogeologic reports and maps of the area, geohydrologic reasoning based on current information, inferences through correlation with known information about hydrologically similar neighboring basins, and drilling pilot holes.

Confined aquifers may have limited connection to surface water bodies or other water-bearing formations. As a result, they have a tendency to be denuded slowly with time. Examples in the Carolinas, South Dakota, southern Arizona, the Ogallala and Edwards aquifers, and many more are outlined in Reilly et al. (2009). Confined aquifers should not be the preferred water source unless other sources are lacking or much information is understood about the recharge rate of the confined aquifers.

Conjunctive Use

The amount of surface water use for municipal and industrial purposes is generally restricted to water that is still available during a 100-year drought ensuring that water is present in ongoing years for such needs. In the same way, many aquifers are managed to produce only the amount of groundwater that is replenished annually (although the recharge amount is clearly exceeded in many basins). In this way, the resources are not depleted and the water needed is available every year. However, demands are continuing to grow while the potential sites for new surface water sources in the United States are dwindling. At the same time, the potential sites that remain face ever-increasing environmental scrutiny with greatly increased costs and environmental mitigation needed if they are to be developed.

As a result of all of the above factors, some utilities are trying to maximize the use of all sources available to them. In the same way that operating multiple reservoirs as a system produces a yield in excess of the firm yield of the individual components, conjunctive use of groundwater and surface water sources can maximize the yield of both under certain circumstances. In addition, at least in the case of confined aquifers, there is often a lesser effect of dry-weather conditions on the aquifer because of the time lag and the relatively slow travel of water in a confined aquifer. Rainfall on the surface of the ground that actually infiltrates into the aquifer is often only a few inches per year.

This is primarily caused by saturation of the aquifer outcrop during rainfall events. Once the pores of the aquifer outcrop (recharge area) are saturated, the remainder of the rainfall runs off and appears as surface water. During dry-weather conditions, the groundwater source is less affected by the dry weather than the surface water source. Where this is the case, over-pumping of the groundwater during 100-year drought conditions may not cause significant reductions in the groundwater table in the short term, but conversion to greater reliance on groundwater may transform to more challenging problems.

As a result, some utilities have investigated means to artificially improve aquifer storage. There are several methods for recharging or filling an aquifer for conjunctive use, both through natural and artificial methods.

Natural infiltration (rehydration). Nature has historically recharged aquifers at a rate that matches natural aquifer drains through losses to the surface through lakes, rivers, and streams. Increased recharge does not occur as a reaction to artificial withdrawals like wells, so water levels can be reduced. Natural infiltration can be boosted by providing additional water supplies through natural infiltration, but artificial infiltration is typically required to help sustain aquifer water levels as described below.

Artificial infiltration. The rate of aquifer recharge and storage can be greatly increased by artificial means, such as through surface spreading and injection. Both methods are described below.

Spreading. Water supply for aquifer storage or recharge can be moved into groundwater aquifers through basins with pervious bottoms. Although recharge occurs through a natural process of infiltration, the process of seepage from the surface into the aquifer, basins store the water on the surface to allow time for percolation from the surface layer and into the aquifer. This method requires significant surface area because infiltration rates vary widely from fractions to tens of feet per day. The federal government and many local and regional utilities have pursued recharge ponds as a means to create sustainable supplies of water. Many are in the West (Washington, Utah, and many utilities in Arizona and California).

Injection. Groundwater injection is often a preferred method for artificial aquifer recharge because the deeper and often more-desirable aquifers can only be reached through injection, and it requires a much smaller footprint than spreading. A cleaner water product, however, is required to limit well clogging, one of three major issues that occur from injection. In some cases, the wells are used solely for injection (e.g., Orange County, California). Others are used for injection and extraction as described below.

Aquifer storage and recovery (ASR) wells. ASR wells are designed to inject water into, and extract water from, the aquifer from the same well. As a result, ASR wells can improve water basin management by storing water underground from periods of excess supply (flood season or wet hydrologic periods) and later allowing a portion of the stored water to be extracted during periods of demand or short supply (such as irrigation season or dry hydrologic periods). There is no evaporation in the aquifer, as with a surface reservoir, and in some ASR wells the recovery is very high. Therefore, ASR wells can be a cost-effective means of aquifer recharge because of increasing land prices. There are other benefits of ASR wells including storage for emergencies; reduction of disinfection by-products, reduction of the potential for land subsidence, and saltwater intrusion; reduction in pumping costs resulting from a higher water table; and opportunities for conjunctive use of surface water and groundwater.

Local conditions dictate success of the ASR concept. Local conditions include geochemistry, bacteria, regulations, and formation type, which can help or hinder ASR efforts. Of the 200 sites that have investigated ASR nationally, only 66 are operational. More than 50 are inactive or abandoned, primarily because of clogging, mobilization of metals (arsenic, manganese, and iron among the most notable), or regulatory issues surrounding the injection of chlorinated drinking water that may contain trihalomethanes (despite numerous studies that demonstrate that the disinfection by-product issues are resolved in the ground).

Because ASR wells have certain unique design and construction features that differ from single-purpose wells, special skills may be required. Using professionals who have an understanding of the concept of injection wells and ASR technology is required to ensure that utility dollars are not misspent. AWWA developed a manual on ASR (M63, AWWA 2015); AWWA and McGraw-Hill also published a book on ASR projects (Bloetscher, Muniz, and Witt 2005); and the AWWA bookstore contains other efforts in this regard.

Successful use of the ASR process requires an aquifer that has a significant amount of storage capacity in the interstitial. Limestone and sandstone appear to be preferred formations. It also requires that there be sufficient transmissivity to allow the water injected into the formation to move away from the injection wells at a sufficient pace to continue to be injected, but not too much transmissivity because the water will tend to migrate away from the well. Where the formation has the appropriate transmissivity, the aquifer should be productive enough to permit extraction of the injected water by the same facilities that treated it. A confounding problem is water with high total dissolved solids. A test program should include permanent monitoring wells and logging of information.

Horizontal wells, infiltration galleries, and riverbank filtration. Infiltration galleries are constructed when subsurface conditions do not permit groundwater development using vertical production wells. Such conditions occur in thin aquifers or where a thin freshwater layer is underlain by saline water. A typical example is a river valley where thin alluvial deposits lie above bedrock where the water is plentiful and the hydraulic conductivities are high, but the aquifer transmissivities are inadequate for well development because of the thinness of the aquifer. In such cases, a subsurface drain or horizontal well is placed in permeable alluvial material and water is collected in a sump connected to a pump. Significant amounts of water can often be extracted through infiltration galleries because of high hydraulic conductivity of alluvial material and their proximity to a recharge source, such as a stream.

Infiltration galleries can either be constructed under the streambed or adjacent to the stream or surface water body. The yield from infiltration galleries beneath a water body is normally twice the yield from galleries adjacent to the water body. Constructing infiltration galleries, however, is usually more difficult under a water body. Infiltration galleries below a water body tend to have higher turbidity and higher bacteria count because of less extensive filtration. Where such withdrawals are considered to directly affect the quantity or quality of surface flows, many of the considerations presented for direct surface water withdrawals will need to be followed, including meeting Surface Water Treatment Rule requirements. This water would likely require filtration, which increases the cost of infiltration galleries (AWWA 2014a).

Many infiltration galleries can be categorized as riverbank filtration (RBF) wells, which are regulated under the recharge Surface Water Treatment Rule rules. It is considered as a pretreatment technology preceding more advanced treatment operations. The concept for RBF wells dates to the 1840s in Scotland and the 1870s in Germany. It has become a common water production technology in western Europe and the midwestern United States. In the United States, RBF systems have been operating for more than 70 years and often provide the only treatment other than chlorination and fluoridation prior to consumption. RBF provides a solution to obtain higher quality water than the river itself (AWWA 2014a).

The concept of riverbank filtration is to use the natural porous sand and gravel sediments underlying water bodies to physically, chemically, and biologically remove contaminants. These contaminants include microbial constituents, metals, organics, turbidity, and other pollutants, which are removed from surface water by inducing water flow through the bottom and/or sides of the water body through pumping groundwater

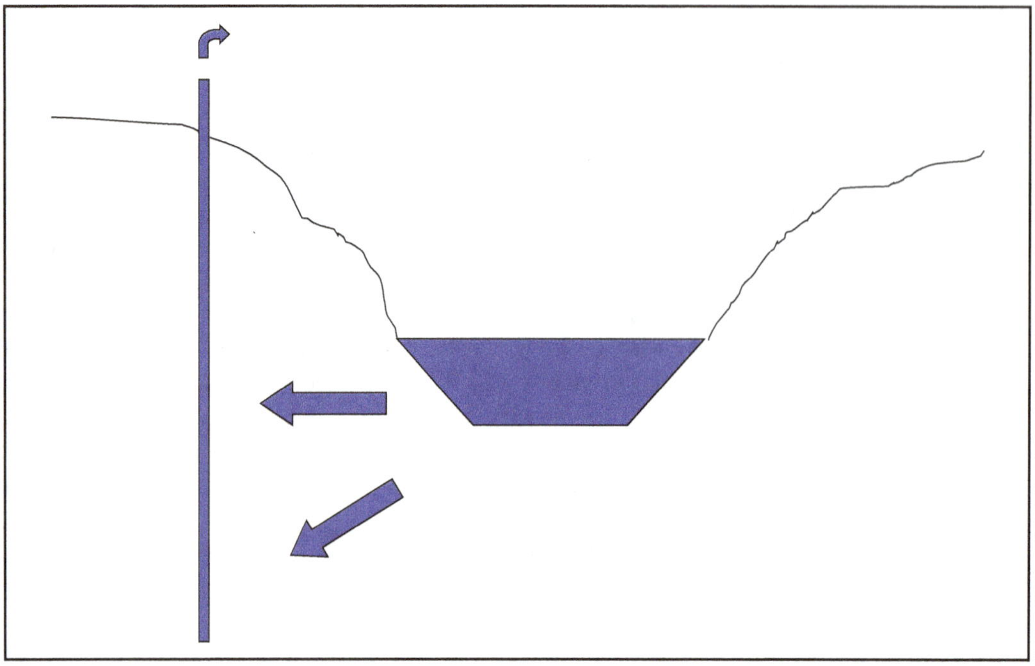

Source: *AWWA 2014a*

Figure 4-1 Riverbank filtration using vertical wells

collection devices (wells). The removal is caused by conditions of induced infiltration created by pumping a vertical or horizontal well. Figures 4-1 and 4-2 show the basic configurations for RBF systems (Figure 4-1 uses a vertical well, while Figure 4-2 shows the horizontal well configuration). More recently in various parts of the United States, RBF has been studied as a pretreatment process prior to conventional or direct filtration and as the primary filtration process combined with disinfection. The idea has been to use RBF as the primary filtration process in lieu of engineered conventional treatment. Because of the ability of RBF wells to filter such contaminants as turbidity and organics, RBF wells are also used in coastal aquifers to naturally filter seawater prior to reaching desalination treatment plants, possibly eliminating the need for pretreatment (AWWA 2014a).

RECLAIMED WATER

Water sources can be garnered from water that has been reclaimed from some prior use. The situation primarily applies to wastewater, but filter backwash and other process waters may also apply. Technically, because there is a finite quantity of water on the planet that has existed for hundreds of millions of years, all water has been previously used at some point and is therefore just being recycled to downstream users.

An analysis of reclaimed water as a source begins with the raw wastewater from which it is derived and ends with its ultimate reuse. For various reasons, in California, the term "recycled water" is preferred as a general label over "reclaimed water," but the reverse is true in Florida (where the term "recycled" is not recognized at all from a water supply perspective) and many other states. A more precise labeling recognizes the terms are related but not really equivalent: water is reclaimed from wastewater to be reused. Thus, the term "reuse water" is also used. For consistency, the term "reclaimed water" is used throughout this manual.

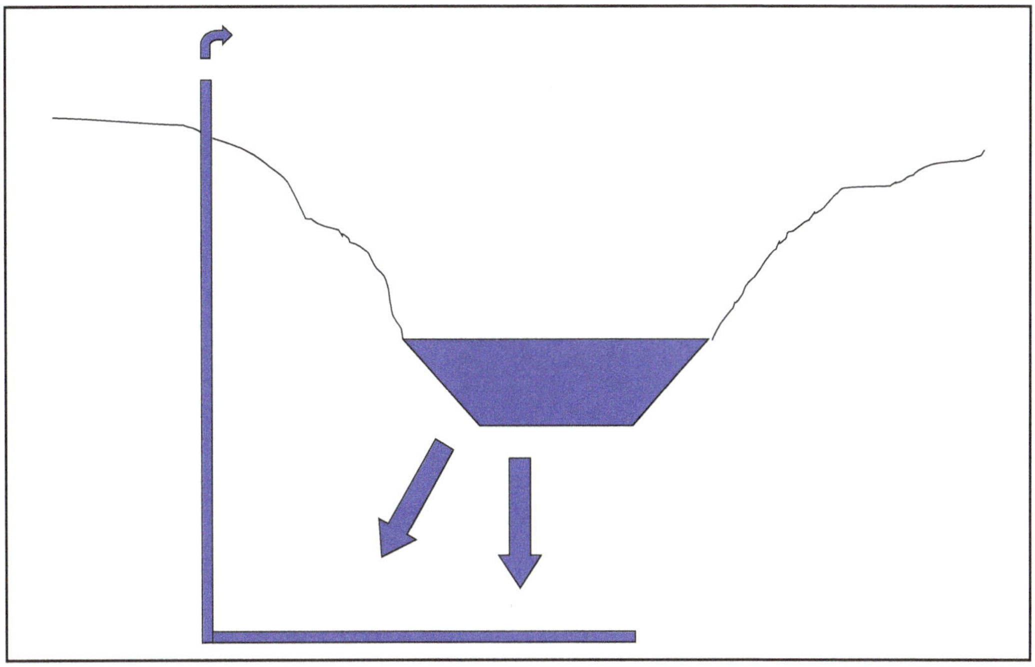

Source: *AWWA 2014a*

Figure 4-2 Riverbank filtration using horizontal wells

The concept of reclaimed water implies intentional reclamation of a nonpotable water source for beneficial reuse by another user, often as a replacement for potable water. In areas with periodic drought conditions or significant water supply limitations (much of the southwest), recovering wastewater for water supplies resolves many challenges. The wastewater plants need to treat wastewater but often lack an ideal place to discharge the treated wastewater. The water suppliers lack sufficient resources to meet irrigation demands. Treating the wastewater (filters and chlorination) and providing a delivery system resolves the concern of both utilities. As a result, reclaimed water programs often provide both water supply and pollution control benefits.

There are two primary means to locate the treatment processes necessary to produce reclaimed water quality effluent:

1. *Centralized treatment plants* collect wastewater from across a service area and treat it at a central location. A distribution system is then required to convey the recycled water to the end user. This method provides lower treatment costs but higher distribution costs.
2. *Decentralized treatment plants,* also called "scalping plants," are located closest to the end user. Although this typically results in higher treatment costs, the benefit is reduced distribution costs.

The following sections describe the different reclaimed water qualities, quantities, and uses. Additional information can be found in AWWA Manual M24, *Planning for the Distribution of Reclaimed Water* (AWWA 2009a).

Water Quality

Municipal wastewater. Municipal wastewater quality across the United States varies significantly but, as a category, has become the primary benchmark for wastewater quality consideration and comparison. Industrial wastewater quality is often characterized by comparison with a quantified municipal standard. Common levels of treatment used for municipal wastewater include pretreatment (usually the type and extent of treatment sufficient to improve a particular raw wastewater quality to that of municipal wastewater), and primary and secondary treatments (generally treating wastewater to an improved level of quality as primarily characterized by removing the major constituents that distinguish municipal wastewater from potable water).

Municipal wastewater is characterized by wastes added during residential or domestic uses of potable water. As constituents in wastewater, these increments include about 200–300 mg/L biochemical oxygen demand (BOD), 200–300 mg/L total suspended solids (TSS), 10–15 mg/L phosphorus, 20–40 mg/L nitrogen, and 200–400 mg/L total dissolved inorganic solids. The most significant other constituent in municipal wastewater, not by mass per unit volume but by potential health effect, is pathogen concentration. This factor is usually expressed by concentration of coliform bacteria, a nonpathogenic indicator organism. For municipal wastewater, the total coliform concentration is about 10^6/100 mL. The following are the typical processes involved in municipal wastewater treatment:

- Pretreatment is required to produce effluent that is substantially higher quality as determined by removing one constituent, such as TSS removal from brewery waste. Alternatively, pretreatment also encompasses removing at least one major constituent that would negatively affect downstream treatment processes, such as a simple grease trap applied to restaurant wastewater. Pretreatment is often a requirement imposed on a wastewater source through a sewerage agency's source control program.
- Primary treatment is the level of treatment required to produce effluent free of most settleable and floatable solids present in raw municipal wastewater. A substantial reduction in total, but not dissolved, oxygen demand is usually accomplished at the same time. Primary treatment is usually provided by the gravity sedimentation process, and its performance is often assessed by the extent of TSS and BOD removal.
- Secondary treatment is the level of treatment required to produce an oxidized effluent in which organic matter has been stabilized, is nonputrescible, and contains dissolved oxygen. Secondary treatment is often characterized by an effluent with TSS and BOD concentrations of 30 mg/L or less. This level of treatment is usually provided by an aerobic biological process, such as activated sludge, performed on good-quality primary effluent.

In the United States, it is required for all wastewater to be treated—at minimum—to secondary quality (85–90 percent removal of organics).

Slow-moving rivers. In slow-moving rivers, advanced (or tertiary) wastewater treatment is normally required to reduce nutrient loads to low levels typically found in the ecosystem. For example, modifying conventional activated sludge to completely nitrify process water and then denitrify it is considered advanced treatment for nitrogen reduction or removal. Advanced treatment is also used to remove a constituent not present in raw wastewater. For example, dechlorination is advanced treatment to remove residual chlorine created by chlorine disinfection. In nitrogen removal, the constituent removed by denitrification is nitrate created by the activated sludge process from organic and ammonia nitrogen present in raw wastewater.

Industrial wastewater. Industrial and commercial wastewater quality is more widely varied than municipal wastewater quality. An absolute characterization of industrial wastewater quality depends on the particular industry; however, a relative

Table 4-1 Example industrial wastewater constituents*

	Concentration (mg/L)	
Industry	BOD	TSS
Auto steam cleaning	1,150	1,250
Wholesale bakery	1,000	600
Hotel with dining	500	600
Industrial laundry	670	680
Market with garbage grinder	800	800
Professional office	130	80
Restaurant	1,000	600
Septage	5,400	12,000

*Data collected and published by the California State Water Resource Control Board.

characterization is made by noting that the concentration of at least one (and usually more than one) constituent is much higher than the municipal wastewater average. For example, waste from a wholesale bakery may have a BOD concentration of 1,000 mg/L, compared with 250 mg/L for municipal wastewater. Table 4-1 lists representative BOD and TSS concentrations for several common industries. Thermal pollution—the degradation of water quality by increasing the ambient temperature—is also a factor to be considered with industrial process wastewater. Much of the industrial wastewater in the United States goes to municipal wastewater plants.

Graywater. Graywater is untreated wastewater originating from showers, bathtubs, clothes washers, and sinks; it is usually of significantly better quality than blackwater, which contains human waste. Graywater systems should be designed so that graywater does not come into contact with humans, rise to the ground surface, or become airborne.

Reclaimed water quality. The amount of treatment needed to allow reuse is generally confined to filtration and high-level disinfection if the water is used for irrigation purposes. This standard is generally termed "advanced secondary treatment." Advanced secondary treatment is the level of treatment required to produce an effluent with less than 5 mg/L TSS and BOD (about 97 percent removal). Advanced secondary effluent is often characterized by turbidities almost as low as those of a potable water supply.

Indirect potable reuse and certain other purposes require much higher levels of treatment. Full advanced treatment includes advanced secondary treatment plus chemical conditioning and reverse osmosis, resulting in virtually distilled water. This water has extremely low levels of pathogens, metals, and most other constituents. It is considered safe to inject in an aquifer for later recovery. For a given wastewater disposal option or water reclamation opportunity, required wastewater treatment levels vary among jurisdictions in the United States, and more so among those around the world. In the United States, individual states promulgate specific requirements. Table 4-2 lists a range of reclaimed water uses and the minimum treatment level required, in order of increased reclaimed water quality. Because of state-to-state regulatory variation, categorizations are illustrative and not precise. As listed in the table footnotes, the level of disinfection varies with reuse requirements. Type of disinfection, as by chlorination or ultraviolet radiation, can also vary with reuse requirements as well as the process water quality being disinfected.

Water Quantity

A critical issue in using reclaimed water is balancing supply and demand over time. If imbalances will occur, storage is generally required.

Supply variations. Wastewater flow generally varies seasonally and with weather conditions. The diurnal fluctuations seen in water systems are generally mitigated in the pipe and through detention time in the treatment process. Seasonal flow variation

Table 4-2 Reclaimed water use by treatment level

Use	Treatment Level
Fodder, fiber, seed irrigation	Primary effluent
Orchard, vineyard irrigation	Primary effluent
Pasture irrigation for milking animals	Disinfected* secondary
Controlled impoundment	Disinfected* secondary
Landscape irrigation with restricted public access	Disinfected* secondary
Food crop surface irrigation	Disinfected* secondary
Uncontrolled impoundment	Advanced wastewater treatment†
Landscape irrigation with unrestricted public access	Advanced wastewater treatment†
Food crop spray irrigation	Advanced wastewater treatment†
In-building toilet and urinal flushing	Advanced wastewater treatment†
Groundwater recharge for nonpotable supply	Advanced wastewater treatment†
Cooling tower supply	Advanced wastewater treatment† plus‡
Groundwater recharge for potable supply	Advanced wastewater treatment† plus§
Live-stream discharge (environmental enhancement)	Advanced wastewater treatment† plus**

*A lower level of disinfection may be allowed.

†A higher level of disinfection may be required.

‡A demineralization process may be required.

§Demineralization and activated carbon processes may be required.

**Nutrient removal and dechlorination processes may be required.

in municipal wastewater generated in a given service area can be substantial, depending on the fluctuations in population (resort communities), weather conditions (wet areas like Florida), and size of populations (smaller communities generally have larger fluctuations). Substantial seasonal variations in industrial discharger wastewater flows is also found, as in canning and certain product manufacturing. Industrial flow variations may be effectively reduced at the downstream treatment plant by a combination of several industries with different seasonality and in combination with flow generation by a larger municipal population base. Diurnal fluctuations are generally limited to small communities.

With industrial wastewater, diurnal variation can be significant (see Figure 4-3). In the extreme, an industrial wastewater hydrograph approaches a step-function distribution: for parts of the day when a plant has no staff and no process generation, there is no flow; for the balance of the day when the process operates, the flow is relatively constant at the maximum rate. A large range of wastewater hydrographs is possible, depending on industrial processes involved, any on-site wastewater reclamation, number of work shifts per day, and other features of process design and staffing.

In larger communities, the pipe travel time, wet-well storage, and equalization of flow in the treatment process eliminates most diurnal variations. Where daily fluctuations are a problem, flow attenuations can be imposed upstream of the plant in the collection system as a predischarge requirement for a larger industrial wastewater discharger, thereby benefiting the entire downstream collection and treatment system.

Additional supply flow variations can result if potable, or some other nonreclaimed supply, is used. Potable water can be used as a short-term contingency to reclaimed water quality problems or quantity shortfalls.

Demand variations. Reclaimed water demand variations depend on use, which include agricultural and landscape irrigation, commercial and industrial processes, environmental applications such as marshes and streams, and others. As with supply, demand may vary seasonally and diurnally.

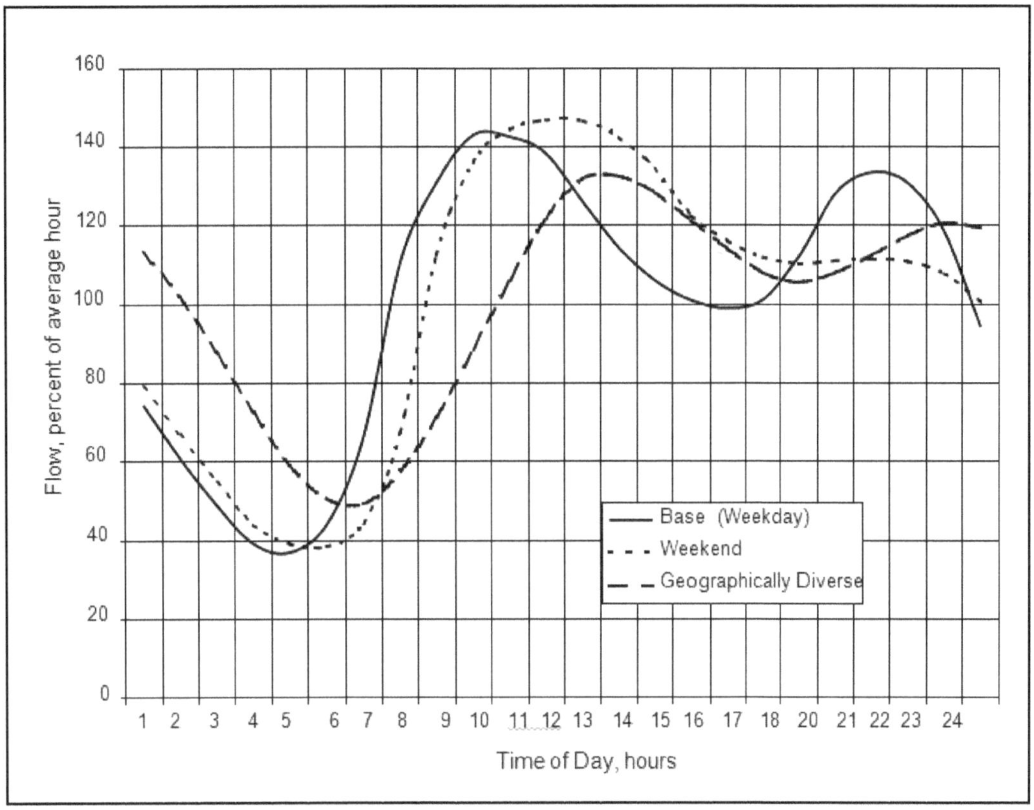

Figure 4-3 Wastewater hydrograph

Seasonal variation in irrigation and some environmental demands are substantial in a given year and from year-to-year depending on prevailing weather. In both cases, the seasonal demand hydrograph tends to complement the associated annual precipitation hydrograph (i.e., irrigation and some environmental augmentation are largely providing needed water when precipitation does not occur). Seasonal variation among commercial and industrial process demands ranges from a little to very substantial. Industries that depend on agricultural production schedules for raw material often have a large seasonal water demand variation. Industries that use raw materials generally available year-round usually exhibit very little seasonal demand variation.

Diurnal variation of reclaimed water demands is often greater than seasonal variation on a percentage basis. Irrigation and many commercial or industrial demand hydrographs have a step-function distribution-off for much of the day, distribution-on for the balance. Most landscape irrigation with reclaimed water is regulated to hours when the public is generally absent from the site, as in a park, schoolyard, golf course, or residential greenbelt. A common daily duty cycle for such demands could be irrigation for 9 hours at night and no irrigation for the remaining 15 hours.

Some uses are buffered from the reclaimed water distribution system with on-site storage. Such storage can be required by the reclaimed water purveyor to control demand variation for the sake of production or distribution facilities, or such storage can be required by the end user for its own purposes. Many golf courses use impoundments, which may double as visual amenities or water hazards, for receiving reclaimed water irrigation supplies during the day and incurring irrigation withdrawal at night. Many commercial and industrial end users also maintain on-site storage tanks for supply-and-demand buffering. Provision for supply-and-demand buffering within reclaimed water transmission and distribution systems is discussed in the next section.

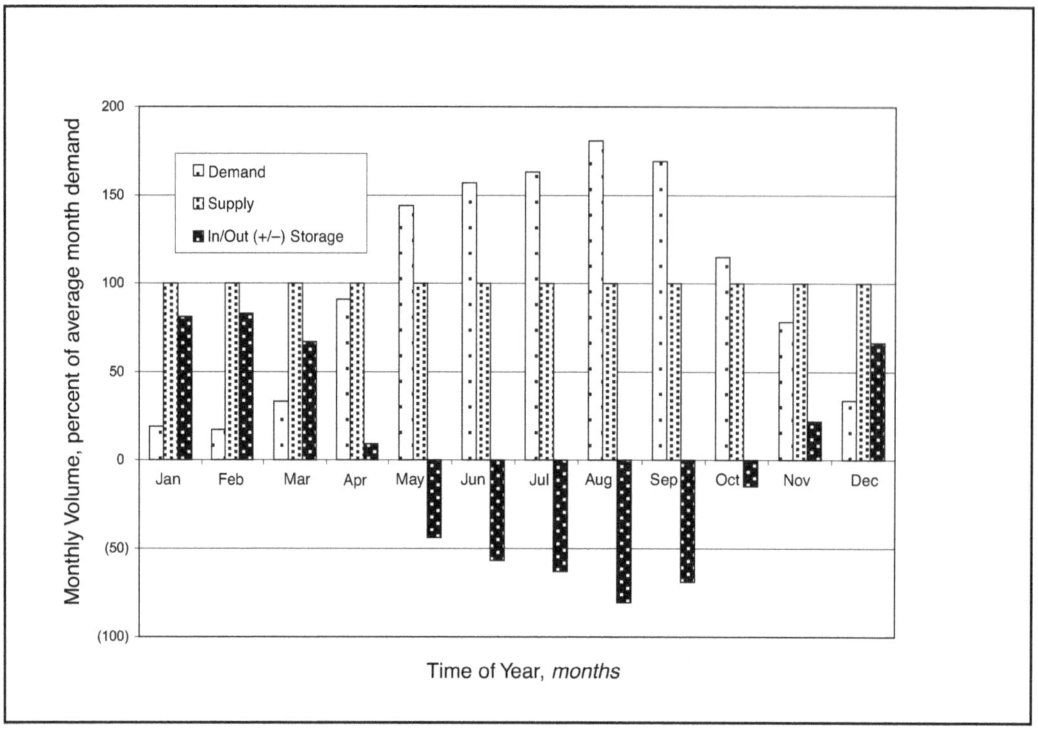

Figure 4-4 Supply sized for average-month demand

Storage. Flow rate differences in the balance of reclaimed water supply and demand over time result in potentially needed storage volumes. As in the discussions for supply and demand, storage evaluations can be segregated into seasonal and diurnal. In all cases, the volume of needed storage centers on particular time dependencies of supply and demand, and includes any defined contingency allotments and dead storage. Lack of adequate storage requires modulating supply production or wasting potential reclaimed water supply and use of nonreclaimed supplemental supply.

The relatively common balancing of reclaimed water supply derived from a relatively large municipality's wastewater and landscape irrigation demand illustrates several points. A seasonal storage hydrograph is calculated as the difference between supply and demand, usually monthly, over the 12 months of the year. If average annual supply and demand are equal, cumulative excess supply equals cumulative supply deficiency, which in turn equals the seasonal storage volume to exactly buffer or balance supply and demand. Typically, several hundred acre-feet of seasonal storage are required for balancing each 1 million gallons per day (mgd) of supply. Figure 4-4 illustrates a typical seasonal storage hydrograph.

Because relatively large storage volumes are needed for relatively small-capacity reclaimed water systems, seasonal storage is most often constructed in open-storage reservoirs. These usually require relatively large land areas and elevation, a dam, and other appurtenances. Because of local disruption created by the dams, such reservoirs are typically no longer constructed. Daily diurnal storage needs are illustrated in the same example. Municipal wastewater flows have large variations throughout the day. Landscape irrigation with reclaimed wastewater is often practiced at night. A diurnal storage hydrograph is calculated as the difference between supply and demand, usually hourly, over the 24 hours of the day. The bimodal supply peak is usually out of phase with the demand step-function distribution, and considerable diurnal storage is required. Typically about 1 million gallons (3,785 m^3) of diurnal storage are required

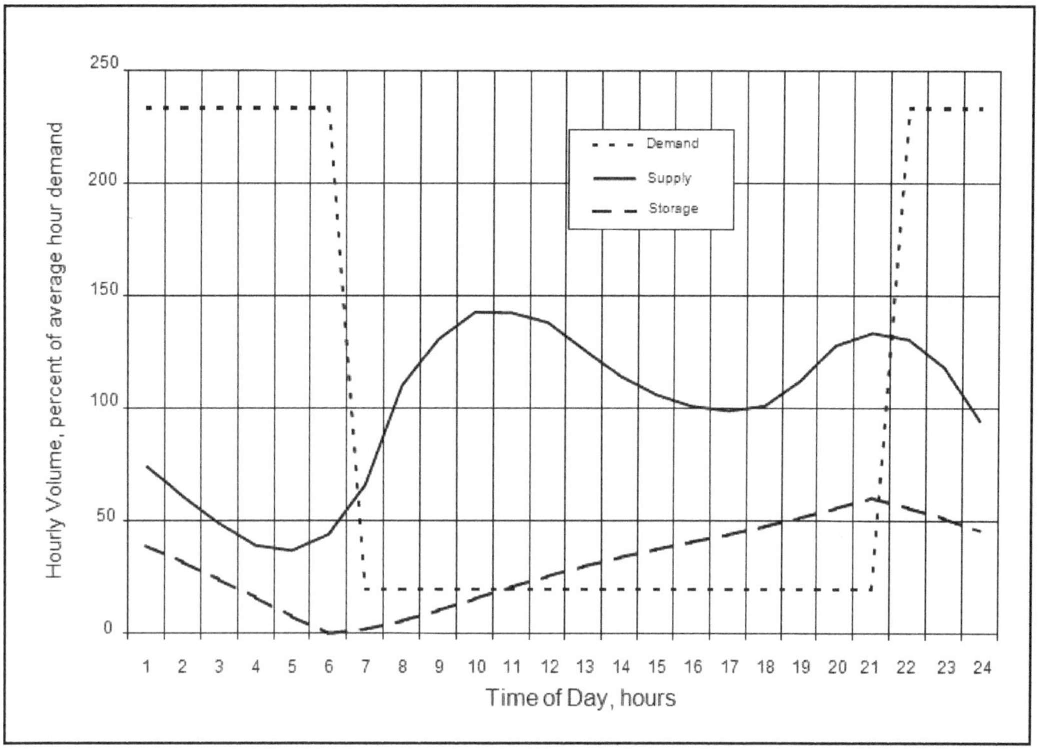

Figure 4-5 Supply sized for daily demand

for each 1 mgd (3,785 m^3/d) of supply. Figure 4-5 illustrates a typical diurnal storage hydrograph.

Reclaimed water diurnal storage is most often provided in aboveground steel reservoirs, on-site irrigation ponds, and sometimes in belowground concrete reservoirs. More information of reclaimed water distribution is available in AWWA Manual M24 (AWWA 2009a).

Uses

Treated wastewater is used throughout the world in a variety of ways. Reclaimed water can be directly applied, indirectly reused, or directly reused by augmenting potable supplies. Each use is described in further detail below.

Direct application. The following are categories of uses for direct application:

- *Landscape irrigation* includes golf courses, schools, parks and recreation centers, athletic fields, freeway medians; landscaped areas around public, residential, commercial, office, and industrial developments; ornamental landscape uses and decorative water features.
- *Agricultural irrigation* historically includes agricultural uses that have represented a significant portion of the total water usage. Benefits for this type of use include the water source reliability and the alternative nutrient source.
- *Industrial* includes cooling water, boiler make-up water, and process water.
- *Urban* includes air conditioning; construction water; fire protection; toilet and urinal flushing; vehicle washing; laundry facilities; window and car washing; and mixing water for pesticides, herbicides, and liquid fertilizer.

- *Environmental and recreation enhancement* uses include creation of marshes and wetlands for wildlife habitat and refuges, recreational lakes, and stream-flow augmentation, although much more treatment is typically needed than for irrigation purposes.

Indirect reuse. Treated wastewater is indirectly reused in many areas of the United States, just by circumstance. Most rivers and streams have substantial components of wastewater because rivers and streams remain the destination for 85 percent of the wastewater disposed of in the United States. The wastewater discharge becomes a part of the river and may be diverted downstream for irrigation or other uses.

Reclaimed water, however, can also be used intentionally for indirect reuse for retarding salt-water intrusion, for downstream capture by water supply wells, or aquifer storage and recovery programs (for irrigation). In the United States, reclaimed water can be used for potable purposes if it meets stringent regulatory requirements including travel through the groundwater aquifer. Many agencies in Southern California have implemented, or are considering, such a program because new dedicated reuse systems are unnecessary if the water can be extracted for potable use.

Direct potable reuse. Direct potable reuse with reclaimed water represents the most rigorous and demanding application and can be accomplished by blending in water supply storage reservoirs or direct input into the water distribution system.

DESALINATION

Historically, desalination processes were used to treat seawater. As more readily available and less costly water sources are no longer available because of increasing potable demands and environmental uses, less desirable water sources are being considered. These less desirable sources tend to have high total dissolved solids (TDS), synthetic contaminants and natural organic matter, and high concentrations of other undesirable minerals. Membranes are also useful for treating these other water sources. The reduction in operating costs is also encouraging communities to consider seawater as a larger component of new water sources. Detailed information is available in AWWA Manual M61, *Desalination of Seawater* (AWWA 2014b).

Saltwater intrusion into coastal aquifers is a problem on both sides of the United States. High TDS water contamination, however, is not limited to coastal areas because upcoming, or upwelling, can contribute to saltwater migration. Within the Colorado River drainage area, as the river flows south to Mexico, agricultural runoff contributes to TDS levels as high as 2,600 ppm at the United States–Mexico border. In fact, the world's largest desalination plant located near Yuma, Arizona, produces up to 72 mgd (0.27 Mm^3/d) of desalted water. The plant was built to treat Colorado River water before it enters Mexico, to meet treaty requirements between the United States and Mexico limiting salinity to approximately 1,000 ppm.

In arid regions of the Southwest, greater water demands are driving water purveyors to spend more money treating lower-quality water supplies, in part to meet these demands and in part because advanced treatment techniques are becoming more cost competitive. Membrane processes have been used throughout the United States since 1962 in Buckeye, Arizona. New water quality regulations are being promulgated for more contaminants, including treatment of protozoa and viruses, color, and organic compounds such as trihalomethanes.

One reason membrane processes are receiving greater interest is that treated water could meet regulatory requirements of microbes and disinfection by-products with fewer treatment unit processes. Highly colored groundwater, due to organic compounds in the

Table 4-3 Desalination process chart

Source Water	Typical TDS Concentration (mg/L)	Variable Process*				
		EDR	RO	MED	MSF	MVC
Potable	1,000	✓	✓	✓	✓	✓
Brackish	7,500	✓	✓	✓	✓	✓
Seawater	35,000	✓	✓	✓	✓	✓
Brine	60,000	No	No	✓	✓	✓

*Electrodialysis reversal (EDR), reverse osmosis (RO), multiple-effect distillation (MED), multiple-stage flash (MSF), and mechanical vapor compression (MVC).

water, can also be treated using membrane processes. Thermal separation processes are used extensively in arid parts of the world and are now becoming more common in the United States. The first US city to use a thermal desalination process to treat seawater was Key West, Florida, in 1967.

Treatment Options

Current desalination techniques fall into two major categories: membrane separation processes such as electrodialysis reversal (EDR) and reverse osmosis (RO) processes; and thermal processes such as multiple-stage flash (MSF) distillation, multiple-effect distillation (MED), and mechanical vapor compression (MVC). Each requires varying degrees of energy, either electric or heat, to convert seawater or brackish water into potable water for municipal use. RO is the preferred process for treating brackish water that typically has a TDS range of 500–10,000 ppm. MSF and RO are the preferred processes for treating saline groundwater, or seawater with a TDS up to 35,000 mg/L. More information on RO can be found in AWWA Manual M46, *Reverse Osmosis and Nanofiltration* (AWWA 2007).

Table 4-3 lists commercially viable desalination processes for typical water TDS concentrations in use worldwide. RO and EDR product water resulting from these processes is referred to as "permeate" because of the process of passing water through the membranes. Approximately 1–2 percent of salt in the source, or feedwater, passes through a membrane process into the permeate. For example, RO can produce 350 ppm product water from 35,000 ppm seawater. The thermal processes MED, MSF, and MVC all produce distilled quality water.

Concentrate Disposal

The major barrier to desalination processes is concentrate disposal. Concentrate disposal costs can be a substantial part of the total cost of desalting and must be considered in the initial planning and design. When using either membrane or thermal separation processes, highly concentrated reject water (concentrate) is produced along with high-quality potable permeate. The TDS of the concentrate depends on the source water characteristics. For example, reject water from a nanofiltration plant may have TDS concentrations lower than 2,000 mg/L. Such relatively low-TDS concentrate can, in some cases, be used for spray irrigation if it is blended with reuse water (for a total TDS less than 500 mg/L). For brackish and seawater treatment systems, reject water concentrations can be significantly higher, typically ranging as high as 30,000 mg/L. Concentrate waste can also typically be characterized by low dissolved oxygen, low pH, ionic imbalances, and may contain corrosion products and residual chemicals from pretreatment processes. Concentrate is also toxic to marine and freshwater organisms because of the ion imbalance issue, which makes surface disposal very difficult. Low-recovery systems also create a large volume of concentrate that is hard to dispose of.

Six common methods for waste concentrate disposal are surface water discharge, indirect ocean discharge from a regional concentrate interceptor to a wastewater treatment plant not producing reclaimed water, land application, injection well, evaporation, and direct ocean discharge through an outfall. One variation on well injection is oilfield flooding. In many cases, sewer discharge may not be an option because of reject water quality or quantity.

DEMAND MANAGEMENT

Demand management strategies include controlling customer demands through maximizing conservation and efficiencies, as well as controlling losses from the supply side through improving supply efficiencies and a successful storage management program. Additional details for demand management can be found in AWWA Manual M52, *Water Conservation Programs—A Planning Manual* (AWWA 2006).

Conservation and Efficiency

Conservation may be viewed as a supplemental or even an alternative technology for meeting safe drinking water needs. Conservation should be implemented as part of a long-term strategy for providing safe and reliable drinking water. One of the chief purposes of conservation is to avoid, postpone, or reduce capital costs associated with new facilities (*Water Conservation Plan Guidelines* [USEPA 1998]). Planners must keep in mind that conservation will reduce revenues, which must be planned for in advance.

Demand side management, water conservation, and water use efficiency are interrelated terms. *Water conservation* has been defined as "the beneficial reduction of water use or water loss" (Baumann, Boland, and Sims 1980) but often gets associated with imposed restrictions on water use. *Water use efficiency* is defined as "using the minimum amount of water to achieve a particular function," or "using water wisely." In general, water conservation, or a change in water use, is achieved through either a change in technology (e.g., changing out a showerhead or an irrigation sprinkler head) or a change in behavior (e.g., taking a shorter shower or resetting the irrigation controller).

As with the adoption of any new technology (or behavior), the change requires three prerequisites: (1) the technology must be available in the marketplace, (2) there must be an incentive (positive or negative) to change, and (3) there must be information available to explain the need for change and how to achieve the change. The technological marketplace for water conservation has changed significantly over the last few decades as witnessed by the competition among manufacturers of toilets, showerheads, irrigation systems, leak detection equipment, and so on. Water conservation has been "incentivized" by the National Energy Policy Act of 1992 and USEPA WaterSense program (2016) that set manufacturing efficiency standards for water fixtures, utility water rates with increasing block rates, and local ordinances that penalize wasteful use of water.

A distinction is made between short-term water conservation measures such as temporary watering restrictions during drought or emergency conditions (discussed in the following section) and long-term conservation measures intent on permanent or long-lasting reductions in water use. In most areas of the country, residential indoor water use has declined because of more efficient water-use fixture standards and building codes. However, in some areas of the country, these gains in indoor water use efficiency have been offset by trends to install multiple-head shower "spas" and automatic lawn irrigation systems. An ongoing debate among water conservation managers is the issue of the "hardening of demand" (or "how low can we go"). Basic water uses such as toilet flushing and showering are generally limited by technical efficiency, while more discretionary uses such as lawn irrigation and car washing can be temporarily suspended during drought or emergency conditions, but these behaviors usually return to normal usage over time. In some areas, the concern is that long-term conservation efforts may have reduced discretionary water uses to the degree that there is little to be "saved" or eliminated with short-term restrictions without adverse economic or social effects.

The benchmark for cost-effective conservation programs is the alternative cost of a new water supply for a given utility. The estimated conservation program costs of administration and implementation can be divided by the anticipated water savings to derive a unit cost (e.g., dollars per 1,000 gallons) of the saved water. As long as this unit cost is less than the cost of a new water supply, the water conservation program is deemed cost-effective. In addition to being cost-effective, a conservation program should target a high water-using end use (i.e., be able to generate appreciable savings) and be implementable from both a social and technological perspective. For example, rain barrels are not effective in areas without summer rainfall, and xeriscape (water-conserving) landscaping may not be widely adapted if homeowners' associations expect green lawns.

Water use efficiency measures at the customer level, such as toilet replacements, high efficiency clothes washers, smart and efficient irrigation systems, and recirculating cooling towers, should be complemented at the system level with 100 percent customer metering, tiered water rates, meter and distribution line replacement programs, and leak detection. Some of the more advanced water efficiency measures such as reuse of nonpotable water for toilet flushing and irrigation may require significant infrastructure or redesign such that they are only cost-effective in new developments.

As discussed in chapter 2, an important component of integrated resource planning is stakeholder involvement. This applies to evaluating water conservation measures as well. Stakeholders should be part of the process of evaluating the effectiveness, acceptability, and implementability of potential conservation measures. The array of potential measures coming out of the stakeholder process should then be appropriate and applicable to the given service area. Furthermore, the stakeholders should be able to see the extent to which reasonable conservation measures can help "fill the gap" between future demand and future supply availability. Many water resources plans often face opposition to the development of additional water supply sources with the argument that water conservation should be able to fill the gap. However, such arguments are often based on levels of conservation and water use efficiency that are unrealistic in many locations. Water leaks within homes, businesses, and water distribution systems will always exist at some level. There is an economic level of leakage, meaning there is a point at which minor leaks are not cost-effective to repair. A finely tuned irrigation system maximizes water efficiency but does not remain so without vigilant management of the system. And many communities are not willing to give up the amenities of lush turf grass landscapes. In some instances, the biggest obstacle to improving water use efficiency is the status quo. Thus, it is through stakeholder involvement in assessing water conservation programs and alternative supply alternatives that the demand management contribution to future water resource solutions can be deemed reasonable and realistic.

It takes significant amounts of energy to move and treat water. Conversely, it takes significant amounts of water to produce energy. This water-energy nexus provides opportunities to consider the collateral implications of water demand management. Studies have been conducted that estimate the inherent energy savings from low-flush toilets and other efficient fixtures. Reduced hot water usage by customers saves on their energy bills, and reduced water demand within the system saves on the utility's energy bills. On the larger scale, reduced energy usage decreases the water requirements of power generating facilities. Such considerations are important when evaluating the sustainability (economic, environmental, and social effects) of alternatives. Demand-side management strategies generally reduce water use and energy requirements, while many supply alternatives such as desalination, reuse, and aquifer storage entail higher energy requirements.

Drought Contingency (Short-Term) Planning

Droughts are naturally occurring adverse components of normal meteorological variations, as are floods, tornadoes, hurricanes, and other storms. *Drought* is an abnormally dry period for a specific area. Drought differs from other adverse meteorological events, such as storms, in several ways. Storm events develop in a matter of minutes for tornadoes, or days for hurricanes. By contrast, the first sign of drought is often a number of beautiful, sunny days. Droughts take months to develop and last for months and even years.

Drought cycles usually contain three phases:

1. First is the meteorological phase or period of below-normal rain. Significant effects have not yet become manifested but will if the dry conditions prevail.
2. The next phase is the agricultural/environmental phase in which soil moisture is depleted and small bodies of water begin to dry up. Crops die, terrestrial plants and animals become stressed, and runoff and aquifer recharge diminish. It is in this phase that most of the economic loss caused by crop damage occurs. Wildfires, dust storms, and abnormal animal movements and death often occur. Water demand increases as landscape and crop irrigation increase, and treatment and distribution systems are often stressed.
3. The third phase is the hydrological phase. Streams begin to dry, lake and aquifer levels fall, springs fail, and water supplies for human needs are stressed.

Precipitation returns to normal; soil moisture is replaced so that normal runoff and aquifer recharge begin to recover; and then streams, lakes, and aquifers return to normal.

What drought management is not. Drought management is often considered the same as water conservation, but the two are different. Drought management involves those strategies needed to temporarily reduce demand and find alternate water resources in response to drought or other conditions that require a quick response. Measures taken can involve those practices normally found in long-term water conservation programs but also include mandatory water use restrictions and rationing in response to drought or other temporary water supply or water treatment and delivery system emergency conditions.

Coordinating drought planning with other water resources planning. A primary function of water resources planning is to provide water for periods of drought. Most municipal water supply plans consider the drought of record when estimating the firm or safe yield of a water source, and most water treatment and distribution facility plans consider peak customer water demand when establishing design capacity. Firm yield calculations for surface water reservoirs and safe yield calculations for groundwater supplies are derived from the available hydrological record. These calculations also depend on the ability to correctly reflect the effects that man has had on that supply when upstream dams, diversions, and return flows are considered, or aquifer recharge and pumpage models developed, to reflect increased pervious cover and other factors.

Therefore, according to the plan, there should be adequate water supply and system capacity to meet increased demand under drought conditions. However, nature does not always operate according to plan. From an integrated water resources planning approach, all demands on the resource, such as agricultural and industrial water uses, and recreational and environmental water needs, must also be considered. As a growing population strains the physical limits of providing water in some areas, shortages will reoccur. In these instances, drought planning may incorporate planned load or demand shedding as a planning tool.

Planning for these contingencies, and determining how much water can be temporarily conserved in times of shortage, is the function of a drought management plan. Three key words—*supply*, *demand*, and *capacity*—are used to describe drought management planning. The purpose of a drought management plan is to manage a period of declining supply and increasing demand so that demand does not exceed either supply or system capacity. It is a way of sharing a shortage or, as one water resources planner put it, "sharing the pain." This is a key difference between drought planning and other water resources planning functions that have as their goal the provision of sufficient water supply and system capacity to meet all demands. More information on developing drought plans can be found in AWWA Manual M60, *Drought Preparedness and Response* (AWWA 2011).

Supply Conservation

Conservation need not only occur on the customer side but can be conducted on the supply side as well. In general, supply conservation involves maximizing the use of existing water resources through methods such as the following:

- Conducting water audits—helps identify locations of significant water loss
- Minimizing leaks using aforementioned water audits and leak detection
- Improving meter accuracy
- Flattening out supply peaks and valleys through storage management as described in the following section

Additional details of supply conservation can be found in AWWA Manual M36, *Water Audits and Leak Detection* (AWWA 2009b).

Storage Management

As previously mentioned, storage management can maximize existing water resources by providing "additional" supplies during peak periods or during times of need, such as during extended droughts. Identifying storage needs and the types of storage are described in the sections below. (Operational storage to meet daily variations is not described in this section.)

Determine storage needs. It is seldom economical, particularly for larger water systems, to provide sufficient production capacity to meet short-term peak demand rates. Determining the most economical combination of production capacity versus storage volume involves several factors, such as the capital and annual operating costs of production and storage facilities, and the manner in which such works can be used to meet system requirements.

Required storage capacity for a domestic system consists of three components: regulatory storage, fire suppression storage, and emergency storage. For most systems, regulatory storage is typically about 25 percent of the maximum daily demand. This allows reservoirs to be used for flow equalization because water fills the storage tanks during periods of low demand and drains during periods of high demand. Fire suppression storage is that volume required to supply the maximum fire flow, and emergency storage is for use in the event of a water supply system failure. There is no particular standard specifying how much emergency storage a water purveyor should have. The amount of storage required depends on available water supplies, inter-connections to other utilities, reliability of power sources, the presence of alternative power sources, and the reliability of the water system as a whole.

Determining the needs for seasonal and/or drought contingency storage is a similar evaluation to emergency storage. It can consist of one or both of the following components:

- *Seasonal storage* consists of the volume needed on an annual basis to balance additional supply during low-demand periods with shortfall in supply during high-demand periods. It is up to the agency to plan for an average year, a peak demand year, a low-supply year sometime in the future or some combination of these.
- *Drought contingency* consists of the volume of storage needed to meet a predetermined drought period.

Storage types. A seasonal storage volume can be obtained both aboveground (surface storage) and belowground (aquifer storage):

- Surface storage is the large quantity of volume needed and typically precludes the use of traditional steel or concrete storage tanks. Storage reservoirs such as those described in the Surface Water section of this chapter are necessary.
- Aquifer storage (conjunctive use) and recovery may be an option in specific local settings to provide the necessary storage volumes.

WATER MARKETING AND TRANSFERS

A tool that provides a supplemental water supply to meet short- and long-term water demands is water marketing. This method includes the transfer, sale, lease, and exchange of water or water entitlements from one water user to another. The methods can be temporary, such as when entitlement to the water is retained but its use is transferred for a short time to another user. In a permanent transfer, water entitlements, gained through water rights or contractual commitments, are sold or transferred to another party.

The general types of water transfers, the issues related to each, and the procedures under which water has been transferred are discussed below.

Types of Transfers

Water has been exchanged among water entitlement holders throughout the United States for many years. Most transfers or exchanges, unless they are directly between adjoining water utilities, have taken place in the western United States. Most were intrabasin transfers, that is, exchanges of water by users situated within the same hydrologic water basin. An example would be one irrigation user transferring water, possibly temporarily, to another when both are being served by the same irrigation water district. In more recent years, water transfers have been developed to convey water from hydrologic basins with surplus supply to basins where supplemental supplies are needed. The intrabasin transfers have been conducted rather informally, with little public scrutiny and regulatory oversight. Interbasin transfers, however, can have significant effects and have been subjected to much more formal and rigorous review and approval procedures. Before discussing the issues related to water transfers, the types of transfers that have occurred and can be used to supplement limited water supplies are defined.

Fallowing. One of the most straightforward ways to create water for sale to others is for a farmer to fallow his ground (not plant a crop for which the water would normally be used to provide irrigation) and sell this increment of water to an urban user. Fallowing frees up water normally used for irrigation in one area for temporary use in another. This type of water transfer is normally used during drought years to provide a short-term supply to an area with restricted supply. In most states, a farmer who is willing to fallow his land must prove to the regulatory agencies that he would have otherwise planted a crop that would have used the water. The proof generally consists of a review of historical farming practices to establish the basis for the type of crop and water use that the farmer had established under similar conditions.

The seller must also be able to prove that entitlement to the water can be incontrovertibly established, and that the quantity of water created represents a reasonable level of consumptive use by the crops that have historically been grown with that water. In this type of transfer, the amount of water diverted for irrigation cannot be transferred for use by others, only the consumptive use portion. The water that would have returned to the system as agricultural runoff from the fields, or through infiltration into the groundwater and then into the surface water system, cannot be transferred. The portion of the original diversion that would have normally been available to other water rights or contractual entitlement holders in the basin remains for their use under normal farming practices.

Crop shifting. Water can also be transferred by shifting to a crop that requires a lower level of water use. This practice is more prevalent with farmers who grow row crops, where a high-water-use crop, such as tomatoes, can be replaced with a low-water-use crop, such as safflower. Normally, this type of transfer is for one year; thus, a high-water-use crop such as alfalfa is not considered because it is more difficult and expensive to substitute. There are many examples of permanent shifts of cropping patterns to create water for transfer, such as shifting farmland used for permanent pasture to a vineyard.

A benefit of crop shifting is that it provides an alternative to fallowing, and the revenue and employment effects on the local area are generally lessened. As with any transfer method, obtaining approval for the amount of water that can actually be transferred to another party requires proof that water is available that would normally have been used.

Groundwater substitution. Many water rights holders also have access to groundwater, and another popular method of water transfer is for the user (usually a farmer, but could be any surface water user) to switch the source of supply from surface water to groundwater. For example, the farmer pumps groundwater to irrigate a crop and transfers an equivalent amount of surface water that would normally have been used for irrigation to another user. A number of issues must be resolved. For example, groundwater pumping may adversely affect others, induce increased losses from the surface stream from which the farmer normally diverts, create overdraft in the groundwater basin, or represent practices that have not normally been used. These factors determine the amount of water that can be transferred.

Groundwater pumping. Another transfer method involving groundwater is pumping water from wells controlled by one user into a surface water body for transfer to others. This type of transfer poses the same questions as groundwater substitution and is very controversial in many areas. In some areas, restrictions on out-of-basin transfers of groundwater have been established. Legal and regulatory restrictions on ground water transfers vary from state to state. Many states have the same types of regulations for groundwater as for surface water rights; groundwater transfers are thus subject to the same restrictions as surface water transfers.

Water conservation. Efficient use of water is important for all users. Because water used for agricultural irrigation in many states represents such a large portion of the surface water supply, conserving agricultural water creates significant supplemental supply without the need for constructing dams or other similar facilities. As the need for water transfers has increased, the potential for selling the conserved water has been viewed by many agricultural users as a method to finance their conservation activities. An example of transfer of water created by conservation practices is the transfer from the Imperial Irrigation District (an agricultural user in the Imperial Valley of California) to the Metropolitan Water District of Southern California and the San Diego County Water Authority (two large urban users in Southern California).

Water conservation practices that can provide water for transfer are very specific to the hydrologic conditions of a given area. Although the need to conserve has been emphasized universally, the conserved water is viewed differently based on geographic location and the fate of the conserved water. In areas where the water lost from unlined canals and infiltration from irrigation enters the groundwater basin or surface streams, the loss may provide a supply for other users. In areas with direct interconnections between surface streams and groundwater basins, the water lost to groundwater may increase downstream surface flows, particularly during summer periods of low surface flow. In these cases, none or only a small portion of the conserved water has been allowed to be transferred.

Stored surface water. Excess reservoir storage can also be transferred. Short-term transfers can be made to provide an emergency drought supply; other transfers provide supply until a new source can be put online temporarily or permanently. The water rights issues related to sale of excess supply that would not have been released except to complete a transfer can be complex, and downstream water rights holders normally object to sale of this water. The withdrawal from storage creates the capability for storage of water in a subsequent year. This water could not have been otherwise stored unless the volume transferred had been withdrawn. The transfer is thus viewed as an adverse effect on their ability to subsequently use the surface water stored in the reservoir. These types of transfers generally apply to water use in the western part of the United Sates and are far less applicable to the eastern states.

Conjunctive use. Another way to create water for transfer is through the conjunctive use of surface water and groundwater to more efficiently manage the joint resource and create more yield. Water banking programs have been created in a number of areas through which excess surface water that is available during normal and wet years is stored in groundwater basins for extraction during years of limited supply. Banking surface water includes in-lieu recharge, surface spreading and infiltration, and groundwater injection. In-lieu recharge programs allow users who normally rely on groundwater to irrigate crops to use surface water provided from another area. By not using the groundwater reserves, surface water is banked and can then be pumped when needed.

Surface water can also be banked by directly recharging the groundwater basin through recharge basins or injection wells. This type of program has become very popular in California, where a number of entities in the Kern County area have established formal programs for banking others' excess surface water, which is available in wet and normal years. These programs are providing a way for holders of water entitlements to better use the yield available during years of excess to provide supplies during restricted supply years.

Related Issues

Water transfers can be very controversial and difficult to accomplish. The major issues that must be dealt with in most water transfers are discussed in this section.

Real or transferable water. As discussed in detail in chapter 3, Legal Issues With Water Supplies, the importance of water to our society initiates significant challenges regarding water rights. Water is considered a public resource, and its use is permitted through water rights and contractual entitlements. Use must be reasonable and beneficial; thus, water that is being considered for transfer must not have been created by unreasonable use or excess to the needs of the water rights holder. In general, the ability to transfer water is based on the need to prove that the water rights or contract holder would otherwise have used the water, and because it was not used, it is available for transfer. Again, this issue generally applies to western United States water rights.

Numerous state and federal agencies are responsible for protecting water resources for public benefit. As a result, water transfers are subjected to a process of review and approval with the primary emphasis on the need to prove that the transfer will not injure other water rights holders or adversely affect fish and wildlife. The amount of water that can be transferred has generally been based on proving the issue of "real water," as opposed to "paper water." In other words, although a holder of water rights or contractual entitlements may have a written (paper) right to the use of the water, determining the transferable water includes examining how much of that water may be transferred. In general, this determination is based on how much water would have been used with and without the water transfer. The "real water" is the difference and generally represents the water that would have been consumptively used by the entitlement holder if the transfer had not occurred.

For example, a holder of water entitlements (water rights or contractual entitlements) may have the right to divert 1,000 acre-ft (1.2×10^6 m^3) for use. Of the 1,000 acre-ft (1.2×10^6 m^3) of diversion, some of this water may have been returning to the surface stream as agricultural runoff. Some may also have infiltrated the groundwater basin, returning as underflow to the surface stream. The real water is the amount that would have been consumptively used by the entitlement holder and would not have been available for use by others. This amount of water can be transferred. Determining this amount is complex and subject to review and approval by many regulatory agencies.

Third-party effects. Movement of water from one area to another through a water transfer can adversely affect the source area unless the transfer is properly planned and managed. Fallowing of cropland is an example of the potential for a transfer to adversely affect a local area's economy. Sales of agricultural products related to growing and harvesting crops normally grown on the fallowed land and the resulting effects on jobs, business incomes, and taxes in the local areas must all be considered. This issue was the prime objection raised to water transfers coordinated by California officials during the droughts of the early 1990s.

Measures that may mitigate third-party effects include limiting the number of acres in a given area that can be taken out of production, restricting the amount of water transferred by a given district or other large holder of water entitlements, and levying a fee on transfers to reimburse those affected for lost revenues or to pay for increased consumer costs. Methods to mitigate third-party effects have been discussed and reviewed in detail in the many public hearings and meetings associated with resolving issues related to movement of water through the Sacramento–San Joaquin Delta in California by means of a joint federal/state planning process called the CALFED Bay Delta Program.

Environmental impacts. A major concern associated with any water transfer or exchange is a potential adverse impact on the environment. Water transfers can degrade surface and groundwater quality, affect aquatic and terrestrial flora and fauna, reduce recreational water uses, cause aesthetic problems, and subside the ground surface, affecting public infrastructure (such as drainage systems and flood control levees in the area of excessive groundwater pumping). Impacts can be local or regional, depending on the size and length of the transfer and the prior base environmental conditions. In general, short-term and intrabasin transfers have relatively minor potential environmental impacts and are therefore subjected to much less environmental scrutiny than long-term and interbasin transfers. Environmental analyses required to facilitate transfers vary among the states and are highly dependent on the environmental compliance requirements of the agencies involved.

Area of origin. Transfer of water from a given hydrologic water basin into another basin can be viewed adversely by many landowners and politicians in the area where the transfer originated. Many objections to the transfer are emotional and related to the desire to retain all resources within the local area for use by its residents. As a result, many local agencies have restricted transfer of water outside their hydrologic basin, unless the seller can prove that the transfer will not adversely affect other local water users or jeopardize local resources. Many landowners have refrained from participating in statewide water transfer efforts during droughts because they fear that participating in such a transfer will jeopardize their water rights or contractual entitlements.

Wheeling transferred water. In many transfers, the water must be wheeled through conveyance facilities (such as pumping stations and canals) that are owned by others, often state or federal agencies. These agencies require that the parties to the transfer obtain an agreement with the agencies to allow use of these facilities to move the water from its origin to the buyer's turnout. Because these facilities have been financed and constructed by the users contracting with these state or federal agencies, they also charge the parties involved with the transfer to use these facilities. These costs can be significant and can affect the economic feasibility of transfers.

By far, the primary factor that has affected water transfers in California is for the parties in a transfer to gain access to these state or federal conveyance facilities. There are many restrictions on pumping water created by concerns over the protection of fish as they migrate through the river system. Because of these restrictions, state and federally operated pumping facilities that export water from the Sacramento–San Joaquin Delta in California do not have the capacity to handle transfer water. In many years, they cannot wheel-transfer water except for very narrow windows of time in the late summer and early fall. As a result, transfer of water from areas with surplus in many years is not possible, if the water must be pumped out of the Delta.

Steps in completing a water transfer. The procedures for developing and implementing a water transfer vary among the states because of the differences in water rights, the conditions of water conveyance in natural and/or human-made facilities, legislative and institutional differences, and the geographical distribution of water supply and areas of need. The length of the transfer and whether it is an intrabasin or interbasin transfer can also increase the complexity of a transfer. Some general steps for implementing a water transfer are described in this section. They generally apply to implementing a long-term transfer, but many are required for any transfer.

1. The first step in any water transfer is to put the buyer and seller in contact with each other to match water needs and available supplies. In some states, clearinghouses where buyers and sellers can make contacts to consider water sales may facilitate transfers. In most cases, however, the buyer and seller are connected through professional water brokers, water rights attorneys, or consulting engineers. A buyer or a seller could also independently identify another party with common interests who might be interested in a potential water transfer by contacting one of the state or federal agencies that has responsibility for water planning, regulation of water rights, or similar responsibilities. Staff in these agencies normally know the professionals who can provide assistance, or they know the buyers or sellers who might share common interests in pursuing a water transfer.
2. Successfully implementing a water transfer requires both the buyer and seller to clearly understand each other's needs and objectives. An efficient way to facilitate the exchange of information is to prepare a memorandum of understanding that defines the basic terms the two parties agree to. This step determines rather quickly whether there are sufficient common interests and objectives.
3. After the parties have generally agreed on the potential terms of a water transfer agreement, the next step is to prepare a simple transfer proposal that describes the basic terms of the transfer and how it will be accomplished. This proposal should describe the amount of water to be transferred, when it will be conveyed from the seller to the buyer, the basis of entitlement by the seller to the water to be transferred, the basis for determining the transfer amount, facilities that will be used to convey the water, and other similar information.

4. After the preliminary proposal has been developed, the parties should meet with the state, federal, and possibly local agencies or individuals that will review and approve the transfer. These agencies and individuals could include the following:
   - State water rights agency, if the transfer requires modification of water rights or approval of differing use of water rights, or downstream water rights holders who may have concerns or objections to the transfer
   - The agency that is a party to the contractual entitlements if the transfer involves use of water obtained through a contract
   - Federal and state fish and wildlife agencies that will review potential impacts or will determine whether, how, and when the water can be conveyed and diverted
   - Local landowners or special interest groups who may have concerns about the effects of the transfer
   - Agencies or individuals who will be involved in permitting conveyance of the transferred water through their facilities
5. The primary purpose of this series of contacts is to describe the basic transfer proposal and to surface any potential concerns that could affect its viability, feasibility, and costs. The need for environmental analyses and documentation, and the scope of any special studies that will be required to support the approval process, can also be defined.
6. Following this initial series of meetings, the parties to the transfer should meet to structure a final transfer proposal that considers the input received when reviewing the preliminary proposal. Either party can withdraw from the agreement if it appears that the transfer will not meet their needs or expectations.
7. If the transfer parties decide to continue, the preliminary transfer proposal should be finalized and submitted for formal review and approval by the state, federal, and local regulatory agencies. This process can vary significantly in length and complexity, and the parties should define the approval process for each agency and the approximate schedule so the use of the water is based on a realistic schedule for approving the transfer. In some states, a transfer that involves using the water for only a year or less can generally be processed quickly, often with minimal environmental documentation. Even with a short-term transfer, however, the primary issue of not damaging legal holders of water rights or adversely affecting fish or wildlife must always be considered. Long-term or permanent transfers can take as long as a year or more if significant environmental documentation or detailed technical studies are required.
8. Environmental requirements vary substantially among the states, and the need for, and scope of, environmental documentation will thus also vary significantly. If formal environmental analyses and documentation are required, the logical approach is to complete them while the review is being finalized by the regulatory agencies. Normally, certification of an environmental document is required as a condition of approving the water rights modifications, wheeling agreements, and other transfer approvals needed to match state and local regulatory requirements.
9. After approving the water rights modifications, completing the environmental process and certifying the documents can take place; any other needed approvals can be obtained, and any needed wheeling agreements can be finalized, and then the transfer of water can begin.

NONTRADITIONAL SOURCES

Several other out-of-the-box sources have been considered over the years but have never gained any significant traction, primarily because of the cost and lack of technological advances. Although not covered in any detail for the aforementioned reasons, these alternatives could become a reality under the right circumstances:

- *Iceberg towing.* Freshwater is stored in icebergs near the earth's poles. It has long been a fairly far-fetched idea that these icebergs could be towed and harnessed for their freshwater and distributed to areas in need.
- *Dehumidification.* There is plenty of water in the air in certain areas of the world. Dehumidification could extract the water from the air in small quantities for use. This concept might best be used in remote areas where centralized water distribution may be difficult.
- *Cloud seeding.* Another concept that has been tested with little success is distributing something in the clouds to cause them to rain in locations best suited for capture and reuse.

REFERENCES

AWWA (American Water Works Association). 2006. Manual M52. *Water Conservation Programs—A Planning Manual.* Denver, Colo.: AWWA.

AWWA (American Water Works Association). 2007. Manual M46. *Reverse Osmosis and Nanofiltration.* Denver, Colo.: AWWA.

AWWA (American Water Works Association). 2009a. Manual M24. *Planning for the Distribution of Reclaimed Water.* Denver, Colo.: AWWA.

AWWA (American Water Works Association). 2009b. Manual M36. *Water Audits and Leak Detection.* Denver, Colo.: AWWA.

AWWA (American Water Works Association). 2011. Manual M60. *Drought Preparedness and Response.* Denver, Colo.: AWWA.

AWWA (American Water Works Association). 2014a. Manual M21. *Groundwater.* Denver, Colo.: AWWA.

AWWA (American Water Works Association). 2014b. Manual M61. *Desalination of Seawater.* Denver, Colo.: AWWA.

AWWA (American Water Works Association). 2015. Manual M63. *Aquifer Storage and Recovery.* Denver, Colo.: AWWA.

Baumann, D.D., J.J. Boland, and J.H. Sims. 1980. The problem of defining water conservation. In *The Cornett Papers.* Victoria, BC: University of Victoria. pp. 125–134.

Bloetscher, F. 2010. The impact of unsustainable ground water. In *Florida Water Resource Conference Proceedings,* May 16–19, 2010. Orlando, Fla.: FWRC.

Bloetscher, F., and A. Muniz. 2008. Water supply in South Florida—The new limitations. In *Sustainable Water Sources (Reno) Conference Proceedings.* Denver, Colo.: AWWA.

Bloetscher, F., A. Muniz, and G.M. Witt. 2005. *Groundwater Injection: Modeling, Risks, and Regulations*. New York: McGraw-Hill.

Hutson, S.S., N.L. Barber, J.F. Kenny, K.S. Linsey, D.S. Lumia, and M.A. Maupin. 2004. *Estimated Use of Water in the United States in 2000*. USGS Circular 1268. Reston, Va.: US Geological Survey. p. 46.

Lisk, B., E. Greenberg, and F. Bloetscher. 2012. *Implementing Renewable Energy at Water Utilities*. Web Report #4424. Denver, Colo.: Water Research Foundation.

Nace, R.L. 1960. *Water Management, Agriculture, and Ground-Water Supplies*. USGS Circular 415. Reston, Va.: US Geological Survey. p. 12.

Reilly, T.E., K.F. Dennehy, W.M. Alley, and W.L. Cunningham. 2009. *Ground-Water Availability in the United States*. USGS Circular 1323. Reston, Va.: US Geological Survey.

USEPA (United States Environmental Protection Agency). 1998. *Water Conservation Plan Guidelines*. EPA-832-D-98-001. Washington, D.C.: USEPA.

USEPA (United States Environmental Protection Agency). 2016. WaterSense home page. https://www3.epa.gov/watersense/ (accessed Aug. 8, 2016).

This page intentionally blank.

Chapter 5

Water Demand Forecasting

Sound water demand forecasts are critical in water resources planning. In particular, the adequacy of supply, optimum facility location and size, and sound transmission and distribution system design depend on the accuracy of demand forecasts. No single method of forecasting will satisfy the varied needs of all water utilities. The forecasting method used and the data needed to correctly apply the method depend on the situation. For example, when a forecast of average annual demand is the primary requirement, a simple per capita approach might be sufficient. When major new capacity must be added in different locations within the utility's service area, a multivariate analysis of disaggregated (i.e., broken down) demand by area and customer class will more accurately specify the size and character of facilities required. These two examples illustrate the interactive and building block nature of the forecasting process. An initial forecast might identify capacity constraints, which in turn prompt more detailed area and seasonal (or peaking) demand forecasts.

It is important to understand the purpose and objectives of the water demand forecast. Is it being developed for regional planning or utility service area planning? The geographic scale and level of detail between these two planning models are different. Will it be used to evaluate water demand sensitivities to changes in water rates? If so, the model will need to include a price variable. The purpose of the water demand forecast is to provide information for good decision making. Thus, it is important to know the priority of the decisions that will be based on the water demand forecast and any specific informational needs of the forecast.

Similarly, it is important to know the study area for which the water demand forecast is being developed. A state-wide or regional forecast is disaggregated into counties or watersheds, while a utility service area forecast may be disaggregated into pressure zones or some other geographic planning unit compatible with a hydraulic distribution model. Thus, the geographic scale of the demand forecast should be determined as part of the initial planning process.

Likewise, the planning horizon (or number of years into the future for which the water demand forecast is being developed) should be determined based on the objectives

Table 5-1 Disaggregation of water demand into sectors and end uses

	Water Uses		
Industry	Potable	Raw	In-Stream
Sectors	Residential Commercial Industrial Irrigation Unmetered System losses	Agricultural Industrial Thermopower Mining	Environmental Hydropower Recreation Navigation
End Uses	Toilets Showers Washing Faucets Landscape Cooling Swimming pools Process water	Crop irrigation Livestock Cooling Boiler feed Process water Extraction Conveyance	Wildlife habitat Power generation Fishing Rafting Barge traffic Downstream flows

of the planning study. The timescale of the water demand forecast should be determined as well. Rarely is a water demand forecast developed on a daily time-step except when peak-day demand is estimated relative to monthly or annual water demand; however, monthly water demands and annual demand estimates are common. Monthly water demand forecasts are often used to assess seasonal variability of demand. Many regions experience seasonal fluctuations in water demand due to climate variations throughout the year. Monthly or seasonal water demand forecasts are often used as the basis for evaluating water efficiency programs, alternative rate structures, and future climate change scenarios.

Water demand can be disaggregated into various components as listed in Table 5-1. A regional water demand forecast may include potable water needs of communities, as well as raw (untreated) water needs of agriculture and large industry, and even considerations of instream water needs. Alternatively, a utility-level water demand forecast usually focuses on the potable water needs of customers with inclusion of estimated system losses, and possibly the volume of potable water needs that can be offset with reclaimed water. Some utility water demand forecasts may provide adequate planning information by estimating the total system demand, while other forecasts should be disaggregated into separate sector, or subsector forecasts, to meet the planning objectives of the study. Some water demand forecasts should be disaggregated to the end use of the water, as indicated in Table 5-1. This level of disaggregation may be necessary when the water demand forecast is intended to serve as the baseline for evaluating water efficiency programs or when the end use disaggregation serves as the basis for building the water demand estimate from the "bottom up."

The general water forecast parameters such as geographic scale, water use sectors and subsectors, time-step, and planning horizon should be decided at the beginning of the development process. These parameters are typically chosen in coordination with other components of the planning process, such as supply concerns. After these parameters have been established, the water demand forecast development process follows these general steps:

1. Collect the data.
2. Analyze the data.
3. Develop a water use profile and water use models.
4. Prepare a baseline forecast.
5. Estimate the effect of conservation on future water demands.

6. Evaluate sensitivity, uncertainty, and alternative scenarios.
7. Update the forecast as new information becomes available.

FORECASTING METHODS

The underlying method of analysis in most forecasting is some form of regression analysis, although it is sometimes not recognized as such. Regression analysis is a statistical technique that mathematically defines the relationship of one dependent variable (e.g., monthly water use) with one or more independent variables that statistically explain the dependent variable in terms of either a fixed or a causal relationship. This process can be applied simply with per capita and extrapolation methods or more rigorously using multiple regression models. The multiple regression models are often used to define or describe relationships between water sales and causal influences, which are then used to estimate future water use under different conditions (i.e., different values of the causal variable in the future years).

Several methods of demand forecasting are often combined, even within a single utility. Different methods may be appropriate for different water use sectors based on the level of data available and level of detail required from the model. Several commonly used methods are discussed in this section, although there is usually significant overlap in their use.

Per Capita Models

As its name suggests, the per capita model simply calculates the total production or consumption per capita for a historical period and applies the current year per capita consumption (sometimes with a trend) to the population projections for future periods. This is the simplest forecasting method and requires only historical production or consumption data, historical population, and a population forecast through the demand forecasting horizon.* The per capita models produce satisfactory results as long as the population forecast is accurate and the customer mix does not change substantially. Using this method, the more critical element in forecast accuracy is usually the population projection, not the per capita water use factor. The obvious exception is when a very heavy-water-using firm enters or leaves the utility's service area, thus changing the per capita rate of use for the area. The primary concern with this approach is a careful definition of the numerator (water use) and the denominator (population). The analyst should understand what level of water use is being used (e.g., total production, billed usage, residential usage) and the demographic area represented by the population. The per capita value may be comparable with equivalent metrics of the same system but may not be comparable with per capita values of other systems.

Extrapolation Models

Extrapolation models plot, in a scatter diagram, annual or monthly consumption related to time or to population and generate a line (with a spreadsheet statistical package) to capture the slope or relationship between the variables. The relationship is then extrapolated into future periods. This process can be applied to total water use or to components of the total, such as residential and nonresidential use. The data requirements are simple: historical production or consumption data, historical population, and projected population.

* Demand forecasting horizon is when the population forecast aligns to the end date of the demand forecast. For example, if the demand forecast needs to project through the year 2030, the population forecast also needs to project through the year 2030.

The relationships to time and population are forms of a single explanatory variable (bivariate) regression analysis without detailed structural analysis, which may decrease their accuracy. The linear form of this model is as follows:

$$D = a + bx + e \tag{5-1}$$

Where:

D = water demand

a = vertical axis intercept term (a constant)

b = derived regression coefficient (slope) of the independent variable

x = independent variable (time or population in the example)

e = error or residual term

The b term is also referred to as the "slope of the relationship between the dependent and independent variables" (i.e., the change in demand [dependent variable] associated with a change in time or population in this illustration [independent variables]).*

The major concern with this approach is the sensitivity of the forecast to the historical data. Extrapolating a long-term forecast from a brief historical period may erroneously reflect temporary or transitory water use patterns, such as the effect of weather or economic conditions. The analyst should understand the characteristics of water use during the historic period and determine if these characteristics are appropriate conditions for the future.

Disaggregate Water Use Models

The billing systems in most large water utilities permit at least some disaggregation of total billings, usually by residential, commercial, industrial, and municipal or public facilities. Further breakdowns are often available for single-family and multifamily residential, institutional, and irrigator accounts. Many utilities can also provide billing data by land-use patterns, region, district, pressure zone, microclimate area, and topography areas. Forecast accuracy is generally improved by disaggregation as long as the water use or trend patterns are different among the segments. If two or more segments are experiencing the same water use and growth rate, separate forecasts are unlikely to improve the accuracy, and additional forecasts are unnecessary. Nevertheless, separate forecasts for residential and commercial accounts are usually made even if their mix is not different or changing.

The basic approach applied to disaggregated sectors of total water consumption is usually to isolate the average *rate of use* per unit, such as gallons per day per unit, or cubic meters per day per unit for each sector; this is then applied to projections of the base units into future years. The *base unit* used could be utility accounts, employment, floor space or acreage, population, or any other unit of measure that makes sense for a particular sector. The base unit also becomes the driver of the forecast for the sector. Thus the future number of units must be able to be estimated or projected. The method of deriving the rate of use per unit can be as simple as taking an average of recent consumption divided by the units (e.g., accounts) from the corresponding time period to arrive at gallons per day per account (e.g., gpd/account, or cubic meters per day per account); or it could be as complex as performing a regression analysis that normalizes the average rate of use per unit for

* Extrapolation models can also take on nonlinear shapes, using polynomial curves (such as $D = a + bx - cx^2$) or logarithmic functions (such as $D = a + b \times \log[x]$) or a power curve (such as $D = a \times x^b$).

weather, month of the year, price, household income, family size, growth, lot size, and conservation effects.

The more detailed analysis can be done using regression analysis of cross-sectional data, which measures the effect of each explanatory variable over a historical time period and across different areas, such as counties, census blocks, or pressure zones. Variables such as median household income and average lot size may account for the cross-sectional differences. These differences would cause a variation in the account per day consumption that cannot be explained by seasonal or weather variations. Consequently, disaggregate water use models are commonly used to isolate and forecast the shifting patterns of component elements within the total demand structure.

Multiple Regression (Multivariate) Models

Multiple regression models simultaneously evaluate a combination of independent variables that can include population, households or dwelling units, household income, lot sizes, land use, employment, and various weather variables. Dummy variables are frequently included to measure the effects of dichotomous (on-off) conditions. A dummy variable is assigned a value of 1 when the condition is "on" and zero in all other periods. For example, the effect of water use restrictions during a drought period during the historical period of analysis could be captured in a regression coefficient of a variable (call it *restriction*) with a value of 1 during the drought period and a value of zero in all other months. This also normalizes the resulting water demand function for such anomalies in the data set. Some forecasters refer to multiple regression models as *econometric models* when the structure of the model primarily addresses economic variables, such as household income, marginal and average water prices, and employment. The basic form of multiple regression models is an extension of the single-variable model:

$$D = a + b_1x_1 + b_2x_2 + \ldots + b_nx_n + e \quad (5\text{-}2)$$

in which the b values are the coefficients (slopes) of the additional x independent variables.

A major concern with this approach is multi-collinearity, which means that two or more of the independent variables are interrelated. If two variables, such as household income and lot size, are correlated with each other, then this interaction will distort the overall model. The analyst should understand the meaning and interpretation of regression statistics and evaluate a series of alternative model configurations to derive the "best fit" model. The analyst should also be cautious of including variables with no causal relationship with water use or variables with low statistical significance. Simply finding a model with a high R-square value does not necessarily indicate an appropriate model.

Land-Use Models

Land-use models concentrate on current and projected uses of residential, commercial, industrial, and public lands within the ultimate boundaries of the water utility. The ultimate boundaries are typically referred to as "build-out" conditions, which are seldom associated with a specific time in the future. Residential land use is frequently divided into two or more density classifications, and the nonresidential categories are also segmented if there are different patterns of water consumption or land development. Growth rates are established (in concert with city planning departments) for each land-use category, usually by census tract, allowing for public land, infill development, and growth into undeveloped areas. Water use factors are developed from historical data for each land-use category (e.g., gallons per day per developed acre) and are then applied to the projections of developed acreage for each category. The concern with this approach is identifying the timing of future growth, when buildout is expected to occur, and how that compares with the utility planning horizon.

Univariate Forecasting Models

Univariate models have limited applicability to water demand forecasting. In univariate or single-variable forecasting methods, only the dependent variable is needed: water production or consumption. This approach offers no explanation of why water demand changes over time and assumes that whatever underlying causes of change in water demand will continue into the future. Two univariate methods are sometimes applied to short-term water forecasting. These methods are the Box–Jenkins, or autoregressive integrated moving average (ARIMA) method, and the method of exponential smoothing. The univariate methods generally are inappropriate for long-run forecasting, but components of these methods (identification of autoregressive patterns, seasonal patterns, and trends) can be useful in developing short-term forecasts. Univariate methods can be used for water demand forecasting, but in virtually all long-term and most short-term situations, a combination of time-series and regression analysis produces superior results. For most utilities, the use of advanced statistical models, which are available for use by analysts with knowledge of statistical methods, is recommended.

Demand Function and Demand Model

Planners will find that using computer spreadsheets makes it easier to construct a data set that structures all the relevant variables in demand forecasts. For the simple per capita model, the spreadsheet must include, at a minimum, historical population and water use. From these data, the planner can calculate the average rate of use (gallons per capita per day or cubic meters per capita per day) to be used in conjunction with future population estimates to derive the future water demand forecast. For the disaggregate model by customer class (and possibly by land use or area), the data set is likely to include multiple years of monthly consumption history for multiple customer sectors or density segments, with weather normalization and perhaps a moving average that tracks cycles and trends. The data set can serve as the basis for estimating the water demand *functions* (i.e., relationships between causal factors and water demand) for each sector.

The water demand function estimated for each customer sector may be in the same spreadsheet or a separate spreadsheet with linked sheets or files that combine with future demographic data and future values of the explanatory variables to generate estimates of future water demand (i.e., the water demand *model*). The water demand model usually contains tabular and graphic output displays. The advantage of developing a fully integrated water demand model is that it allows planners to judge the relevance and potential forecast accuracy during the building process. The disadvantage, especially in a highly complex utility, is that developing a fully integrated model can require a high level of statistical skill and much effort. Thus, many utilities engage consultants to conduct part or all of the demand forecasting process.

SELECTING THE RIGHT MODEL

Selecting the right water demand forecasting approach is based on three criteria: meeting the goals and objectives of the study, data availability, and budget. First, regarding the water demand forecast, it is important to understand its purpose and objectives and to know the level of decisions that will be based on it. The demand forecast must be able to generate the information needed by planners and decision makers. There is no point in developing a complex water demand forecast when a simpler model will do. Alternatively,

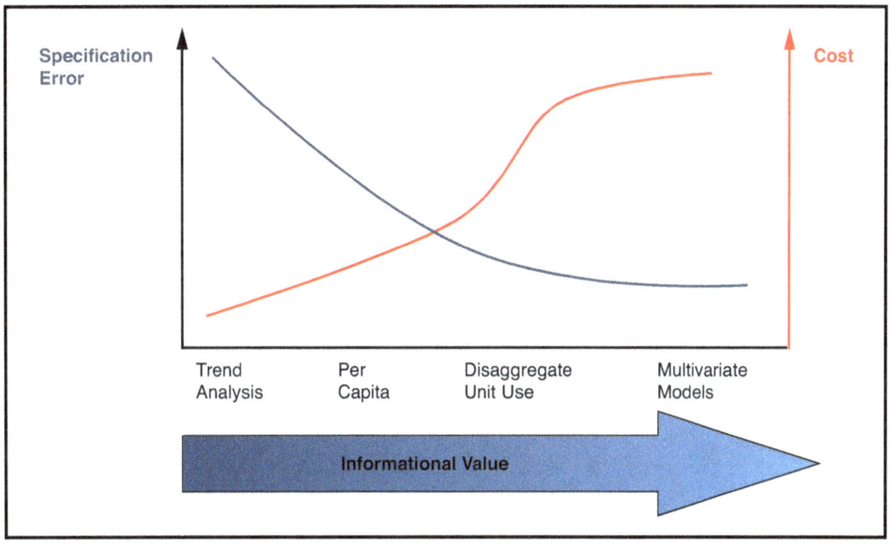

Figure 5-1 The specification error–information–cost trade-off

a simple demand model such as a per capita model does not answer complex questions such as the effect of future rate changes, weather changes, or alternative growth patterns.

Second, planners need to determine what data are available, how reliable it is, and what type of models can be developed from the data. Data availability and reliability may limit the complexity of model development. If additional data are needed beyond what is readily available, can the data be obtained? A list of elements commonly collected for developing a water demand forecast is included at the end of the chapter. The availability and complexity of data may limit the form and structure of the demand model. In addition, the availability of projected data (the future units for each sector and future values of explanatory variables) must be considered. Data should be obtained from reliable sources. For example, population projections might be obtained from a regional planning agency that all stakeholders agree with. A water demand forecast based on data from unreliable sources may not be deemed credible when it is reviewed by others.

Finally, the development of a water demand model must be within project budgets and schedules. There are financial and time constraints to data collection, data analysis, and model development that must be considered. More-complex models require more time and effort to develop than less-complex models; however, the trade-off is that more-complex models are more informative and offer better decision-making support, as illustrated in Figure 5-1. Also shown in Figure 5-1 is the trade-off between model specification error and model complexity. The more statistically complex models offer less specification error along with more informative value, but at greater cost.

Selection of the right water demand model also depends on the level of disaggregation of the water demand forecast that is being developed. For example, a per capita water use estimate can be developed for an entire utility service area in which the estimate represents total water production for the system. Alternatively, a per capita estimate can be developed for each customer sector within the service area. More disaggregate models require more data, which may mean that additional base unit and potential explanatory variables must be identified for each sector, as illustrated in Table 5-2. Final selection of the explanatory variables of each sector demand model is dependent on the outcome of statistical analysis of the data for that sector. Disaggregate water demand models that estimate water demands for different customer sectors must also include estimates of system losses, which are not incorporated into sector demands but are part of the total system water demand (production).

Table 5-2 Possible water use sectors, associated units, and explanatory variables

Sector	Unit	Variables
Residential Residential, single-family Residential, multifamily	Use per day per account Use per day per household Use per day per person	Household income Person per household Density (units per acre) Temperature Precipitation Marginal price of water Average water bill
Commercial Industrial Institutional Public/Municipal	Use per day per account Use per day per employee Use per day per acre	Income Production output Employment mix Labor productivity Unemployment rate Temperature Precipitation Marginal price of water Average water bill

ANALYSIS OF CURRENT WATER DEMAND, FACTORS, AND TRENDS

Once historical water use data has been collected, it can be analyzed to identify water use patterns and trends. It is useful to create a "water use profile" of the study area from the available historical data. Such a profile should characterize total water use (production) and water use among each of the identified water use sectors. Historical trends in water use, as illustrated in Figure 5-2, can be helpful in identifying trends, understanding factors that affect water use, and patterns within these factors. The data shown in Figure 5-2 indicates that retail water use no longer increases proportionally with the number of customers. This suggests changes in per capita water use, possible shift in the residential/nonresidential mix of customers, and the influence of pricing and water conservation on total consumption.

Tables and graphics can be developed that show the proportion of total water production that is billed by each customer class and the proportion that is nonrevenue water. Graphs of these data can reveal trends that can also be determined through statistical analysis of the historical data in conjunction with concurrent historical weather and socioeconomic data. The seasonality (monthly distribution) of water use by sector is important, particularly in areas where seasonal water use is driven by outdoor irrigation and cooling uses.

An average rate of use per unit should be calculated for each water use sector. Regression analysis can be used to identify factors that influence changes in the average rate of use over time. Regression analysis can also generate information that can be used to normalize the average rate of use for historical variations in weather, the economy (recession), rate changes, and conservation. For example, binary variables representing the months of the year or a seasonality index, as shown in Figure 5-3, can be used to separate weather effects from normal seasonal variations. With simpler water demand models, an average rate of use should be derived from a period of time that is relatively free of abnormal weather conditions and other unusual events.

Identifying indoor and outdoor consumption is critical in evaluating price and conservation effects on water use. Indoor and outdoor water use can be estimated with acceptable accuracy from the historical monthly water use data for each sector. The recommended method for this task is the following:

1. Calculate the average water use for each month from the historical data. For example, average all the monthly water use for all Januarys in the data set, then Februarys, and so on.

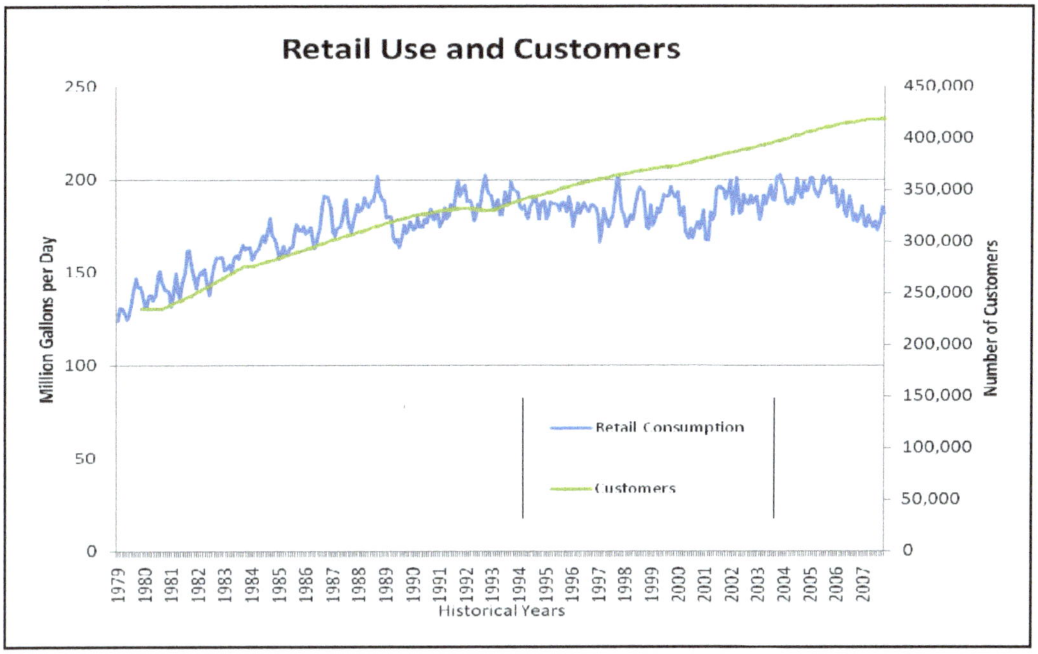

Figure 5-2 Historical trends in water use

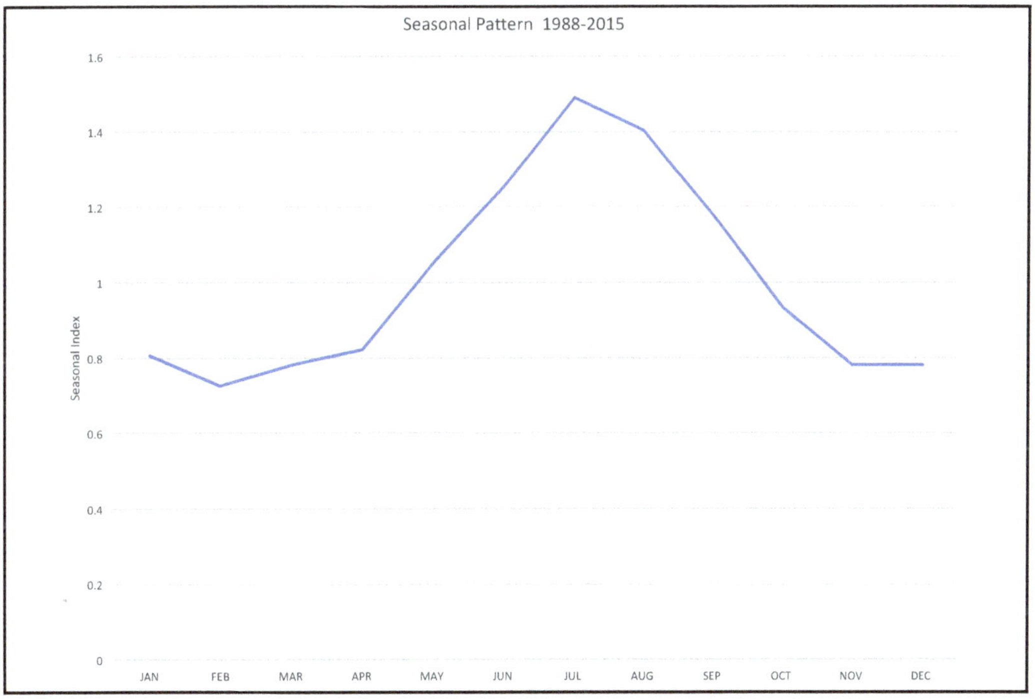

Figure 5-3 Seasonality index of single-family water use

2. Select the month with the lowest average monthly water use or seasonal index as a proxy for indoor water use. This estimate of indoor use is fairly reliable for areas that have a cold winter climate with no irrigation. For warm weather areas that have irrigation during winter months, this estimate may need to be adjusted for the proportion of monthly use that is irrigation.
3. Subtract the estimated indoor water use from average water use of each month to estimate outdoor water use by month. The total indoor use estimate across all

12 months and the total of outdoor use across all months can be used to derive the percentages of total annual water use that are indoor and outdoor water use.

Although they vary by customer sector, the proportions of indoor and outdoor water use are likely to remain the same into the future unless significant water efficiency programs reduce either of them.

Water use does not usually decrease when water prices are increased reasonably in line with inflation, that is, when there is no significant increase in the *real* cost of water. However, large increases implemented either to provide funds for major operating or capital projects or to induce lower consumption levels are likely to have statistically measurable effects. Numerous studies of price responsiveness (elasticity) are available in water industry literature, which should be reviewed for methods and unique situations.

Price elasticity is defined as the percentage change in consumption divided by the percentage change in price and can be expressed in various formula formats. One of the most practical expressions is

$$E = (dq/dp) \times (p_i/q_i) \quad (5\text{-}3)$$

Where:

E = elasticity

dq = change in quantity consumed

dp = change in price

dq/dp = regression coefficient of price (the slope of the relationship)

p_i = price at some point on the demand curve (e.g., average of the price values)

q_i = quantity consumed at the same point on the demand curve (e.g., average of the quantity values)

Note that price elasticity, or other elasticity measures, depend on the functional form of the demand model. For instance, in the case of a double log functional form, the coefficient of the price variable is the price elasticity.

The purpose of the elasticity analysis is to identify the magnitude of change that occurred in the past or might occur in the future under various conditions. For example, elasticity analysis allows forecasters to estimate the effect that a large rate increase has had in recovering the capital cost of new treatment facilities or to measure the effect of drought rates on water use during a period of supply crisis. It can also help forecasters estimate the effects of implementing conservation rates to reduce water use over an extended period. The price response may be affected by the history of rate increases. If rates have been repeatedly increased, the demand could be *hardening* (more inelastic) because the discretionary water uses have already been eliminated. More information on water rate effects and price elasticities can be found in AWWA Manual M1, *Principles of Water Rates, Fees, and Charges* (AWWA 2012).

Similar to the analysis of price elasticity, elasticities can be calculated for changes in income, weather, and other variables in the demand model. Elasticities with respect to changes in temperature and precipitation can be derived from historical monthly data and are used to evaluate future scenarios in which average monthly temperatures and precipitation change from the historical averages.

DEVELOPING A FORECAST OF FUTURE WATER DEMAND

Once data have been collected and analyzed, and current water use patterns have been identified, the next step is developing the model for estimating future water demands.

Forecasting models are typically created in a spreadsheet format such as Microsoft Excel. The basic model structure depends on elements such as the number of sectors, geographic disaggregation, types of demand functions, planning horizon, and alternative scenarios of the demand forecast. Simpler models may include both historical data and projections, while more complex models may have separate files for historical data with data analysis and the water demand forecast model. A typical water demand model spreadsheet may have separate worksheets for each of the following elements:

- *Introduction*—a description of the service area, who developed the model, the date and version of the model, and other identifying information
- *Demand functions*—identification of dependent and independent variables, with statistical coefficients for each variable for each customer sector, as well as the demographic drivers (units) for each sector
- *Input data*—input values for each of the independent variables and demographic drivers identified by the demand functions; these will vary by sector and time period, and may include variable inputs for alternative scenarios
- *Base period forecast*—an estimate of water demand by sector for a specified current or historical period of time that can be compared with actual water production or consumption data for verification and calibration of the water demand model
- *Baseline forecast*—an estimate of water demand by sector for the full planning horizon
- *Alternative forecasts*—variations from the baseline forecasts given alternative demand function inputs (such as alternative growth, drought, or weather scenarios) and conservation effects
- *Summary tables*—a set of summary tables comparing the baseline and alternative forecasts in a printable format
- *Graphics*—a set of charts and graphs illustrating the baseline and alternative forecasts in printable format

Once the forecast model has been created, it should be tested. Standard quality assurance procedures should be followed for testing and verifying spreadsheet formulas and calculations. In addition, the demand forecast model estimate of water demands in a base period should be verified. A base period should be carefully selected for model verification. It may be necessary to use a multiyear average as the base period to avoid the variation in water demand in any given year because of weather, rate changes, conservation, and other influencing factors. Input data for the base period should correspond with the time period of the observed water volumes to be used for model verification. In some instances, forecasters will input historical data into the water demand model to create a "backcast" of water demand in historical years for comparison with actual data. In such cases, a backcast of multiple years can be developed to trace the variation in water demand that results from actual conditions from year to year. It is important to avoid comparing "apples to oranges" when comparing the model estimate of base period water demand with actual data. Differences in data sources, geographic scale, categories and sectors, nonrevenue water, effects of conservation, rates, economy, and other influencing factors may have to be adjusted before valid comparisons can be made.

The baseline water demand forecast of a simple demand model may be an estimate of total water demand or total water production. More detailed water demand models will have individual estimates of water demand by sector, such as residential, commercial, and industrial water consumption. In addition to the estimates of water demand by sector, any model of a utility system should include an estimate of nonrevenue water. *Nonrevenue water* (NRW) is usually defined as the difference between total water production and total metered water consumption. NRW is composed of three major components, as identified by the International Water Association and described in detail in AWWA Manual M36, *Water Audits and Loss Control Programs* (AWWA 2016). These three components are unmetered water use, apparent water loss, and real water loss. Unmetered and unbilled water use is water that is actually being used by customers but not being accounted for in the billing data. This portion of NRW is generally small in the United States where most utilities are moving toward 100 percent metering of water use and automated billing systems. Apparent water loss represents the portion of NRW that occurs because of unauthorized uses, metering errors, and billing errors. Apparent loss generally does not represent physical water usage (except for unauthorized water use). Real water loss is the physical loss of water due to reported and unreported line breaks, reservoir overflows, and leakage within the distribution system. Most utilities have water loss control programs of timely break repair, and ongoing leak detection and line replacement, in an effort to minimize real losses. These components of nonrevenue water should be understood and properly represented in the water demand model when comparing estimated water demand with actual water use data.

Typically, the verification process using the base period comparison will lead to a calibration of the water demand model. If the verification results in a difference greater than 5 percent, then a calibration of the model is in order. If the demand model has separate functions for separate sectors, then each sector can be calibrated as long as there is an "apples to apples" comparison.

Once the base period demand estimate has been verified and the model calibrated, the water demand model is ready to be used to estimate future water demand. Projections of future demographics, (i.e., the units that drive the forecast) should be obtained from a consistent and acceptable source, such as a regional planning authority. The baseline forecast—that is, the forecast for the planning horizon for "normal" future conditions of weather, current price and economic conditions, and current levels of conservation—can be estimated. Typically, a stakeholder review of demographic projections and the initial baseline forecast is used to determine the acceptability of the forecast and its underlying data and assumptions before the forecast is finalized or can be used to assess alternative planning scenarios.

The baseline water demand forecast is the benchmark for evaluating the effects of future alternative demand scenarios. The most common alternative forecast is that of planned and potential future water conservation programs. Potential future water conservation effects can be estimated in one of two ways. The first method is to estimate the percentage reduction by sector from specific water efficiency measures. The second method is to develop estimates of potential water savings from an end-use model in which sector water demands are disaggregated into individual end uses and the effects of potential water efficiency measures on individual end uses within each sector are estimated. Such estimates are often based on literature reviews and manufacturer's estimates of potential savings of new water efficiency technologies. More detail on estimating potential effects of water efficiency measures is found in AWWA Manual M52, *Water Conservation Programs—A Planning Manual* (AWWA 2006). Estimates of future water savings from conservation measures are subtracted from the baseline estimates of sector water demand for each future time period. Depending on the sophistication and detail of the water demand forecast and the estimated savings, conservation savings can be subtracted from sector estimates of indoor and outdoor water demands as appropriate. Once the baseline water demand forecast is adjusted for potential conservation savings, the estimate of NRW should be recalculated. In addition, programs intended to improve (reduce) water loss can be represented in a decreasing percentage of NRW (as a percentage of total production) over time within the demand model.

The baseline and conservation water demand forecasts are two common future water demand scenarios. Other future planning scenarios can be developed that assess alternative demographic growth conditions, future climate conditions (climate change), and potential rate changes. In addition, uncertainty of the water demand forecast can be evaluated through different means.

EVALUATING UNCERTAINTY OF THE WATER DEMAND FORECAST

All forecasts will include some unavoidable margin of error. If the total demand forecast is carefully constructed, the volatile elements can be isolated and tested for their simple and joint sensitivity in the total forecast picture. The elements of permanent variation compared to temporary variation must be isolated. For example, a sensitivity analysis of water demand among single-family residential accounts might examine variation in key variables such as the projected number of households, the rate of use per account (gpd/a, or m^3pd/a) and the effects from naturally occurring and programmed conservation measures. These components will have an effect on demand. Sensitivity of demand in any given year (or month) can also be evaluated for variations in the deviations of weather from normal.

Variables that are most sensitive to error in the long-term demand forecasting process must be identified. Usually, long-term forecasts of population and employment, or other underlying demographic drivers of the water demand forecast, generate more sensitivity than the estimated rate of use (gpd, or m^3pd) per dwelling unit, per employee, or per population. Several methods can be employed to define the variation related to the use of selected variables.

Confidence Intervals From the Standard Error of the Estimate

The least-squares method of deriving regression coefficients identifies a regression line through a two-dimensional or larger array that minimizes the squared deviations of actual values of the dependent variable from the fitted values. If the model is properly specified and unbiased, the residuals (forecast errors) will be normally distributed around the regression line. The measure of dispersion that results from the regression analysis is called the *standard error of the estimate*, which is similar to the standard deviation that can be calculated for any single variable. The usefulness of the standard error is that probability theory can then be employed to define *confidence intervals* around the regression line for the period of fitted data and for a forecast period. The standard error reflects a type of joint or conditional probability in the sense that the probability distribution is around an estimate of the dependent variable that was derived from the input of any number of significant independent variables in the regression equation. Thus, the forecast can be presented with expected values and confidence intervals (above and below the expected values of the forecast) representing a defined level of statistical significance.

Standard Errors of Coefficients

Each independent variable in a regression equation has a calculated coefficient, significantly different from zero, that expresses its effect on the dependent variable for the addition of each unit of that independent variable. A standard deviation can be calculated and applied to each variable coefficient to define a probability range around the coefficient value. Some variables that are highly significant in predicting water demand will have a narrow probability distribution around their coefficients. Other variables that are not as strongly identified in predicting water demand will have a much wider probability distribution around their coefficients. The magnitude of the coefficients multiplied by the variable values, combined with their dispersion, defines the importance and volatility of the variable in demand forecasting.

Joint Probability and Monte Carlo Analysis

Each variable in a regression equation has a mean value and a probability distribution around it. All elements of the forecast (model parameters and inputs) are assumed to be independent of one another. To test the forecast sensitivity, it is possible to combine the dispersion (probability distribution) of different elements of the forecast, such as the per-unit factor (e.g., g/d/a) with the dispersion of the unit forecast (account growth). A practical method to define the probability distribution around the key variables of the model is to estimate the high and low possible value of each variable. The Monte Carlo simulation process can be added to a spreadsheet version of a water demand forecast using commercially available statistical add-on packages. A Monte Carlo simulation randomly selects a value for each model parameter and model input from the possible range of values for each element. The forecast resulting from the selected elements is recorded and the process is repeated for thousands of iterations. The result of the analysis is a joint probability distribution of the forecast that reflects less variation than a distribution derived from the extreme values of the independent variables. The inference is that, in facility planning, one should not plan for the "worst-worst" case, which would be made up of the extreme adverse effect of each variable occurring at the same time. The joint probability distribution allows for offsetting (high and low) occurrences among the set of variable combinations. The baseline water demand forecast is typically the mean (50th percentile) of the probabilistic forecast.

Scenario Analysis

As indicated earlier, a spreadsheet format of a water demand model gives the planner flexibility in creating alternative scenarios. Model inputs such as demographic growth rates, future weather conditions, future water pricing, and levels of conservation implementation can be varied to create alternative water demand forecasts. Unlike the joint probability analysis, the scenario analysis does not create estimates of the likelihood (probability) of the outcomes. However, scenarios can provide planners with a range of potential outcomes that bracket the baseline water demand forecast and can be generated with less effort than the probabilistic forecast.

SUMMARY

Water demand forecasting can be a very technical and intensive process in certain situations. The first rule should be to evaluate the degree of disaggregation and complexity necessary to provide the required accuracy and information needed from the forecast. If the planner cannot easily access the data for disaggregated sectors and areas or if the accuracy of the data is suspect, the planner should try an 80 percent solution. The forecasting process will differ depending on each agency's needs and analytical abilities.

To some degree, all forecasts are inherently flawed. The key to efficient forecasting is to identify the important variables for the type of forecast prepared. Once a forecasting model has been properly specified, it is far more efficient to recognize the inherent error level and work with it to identify upper and lower bounds of potential water demand than to try to fine-tune a model to achieve the last possible percentage of efficiency from a limited database.

Finally, an important step in the demand forecasting process is to periodically review and update the forecast. New data, such as demographic projects, should be incorporated into the water demand forecast as available. New building codes and manufacturing standards regarding water use fixtures affect the rate of water use over time. Land-use patterns, local ordinances, and growth patterns can change the outcome of water demand projections; thus, it is prudent to periodically update and refine the water demand forecast.

REFERENCES

AWWA (American Water Works Association). 2006. AWWA Manual M52. *Water Conservation Programs—A Planning Manual*. Denver, Colo.: AWWA.

AWWA (American Water Works Association). 2016. AWWA Manual M36. *Water Audits and Loss Control Programs*. Denver, Colo.: AWWA.

AWWA (American Water Works Association). 2012. AWWA Manual M1. *Principles of Water Rates, Fees, and Charges*. Denver, Colo.: AWWA.

This page intentionally blank.

Chapter 6

Water Quality

When planning water resources, the quality of potential sources is as important as the quantity of water each may provide. Ideally, source water should be free of microbial contamination from human and animal activities, free of toxic natural and synthetic chemical contaminants, contain limited nutrients, and be of adequate quantity to meet the needs of the population served with as little treatment as needed. The quality of water sources ranges from nearly pristine in some areas to very contaminated in others.

Pristine sources are rarely available. Increasing populations, increasing pollution from domestic and industrial activities, and other pressures cause many communities to use inferior quality source water. The poorer the source water quality, the greater the need to provide multiple-step treatment to remove microbial and chemical contaminants. Each increment of treatment increases the cost of the finished water.

This chapter introduces the main concepts of water quality and relates them to water resources planning. Topics covered include the hydrologic cycle and water quality; specific quality issues for surface water and groundwater sources; and the physical, chemical, and biological components influencing water quality. It also discusses the sources of water contaminant loading, and water quality sampling and monitoring.

THE HYDROLOGIC CYCLE AND WATER QUALITY

Water supply sources are primarily surface water or groundwater. A region's surface water and groundwater resources are closely interconnected by the hydrologic cycle. Many potential contaminants, whether natural or human-made, affect the quality of water supply sources. These are further detailed in chapter 7. These contaminants can enter a water resource at any point in the hydrologic cycle.

The hydrologic cycle is the continuous movement of water through the environment, above, on, and below the earth's surface. When water evaporates, moisture forms clouds and then falls in the form of precipitation. The precipitation that does not evaporate falls onto land or surface water, or is intercepted by plants or structures. The water flows across the land or stays in surface water until it evaporates or percolates into the ground. Trees and other plants also take up water through their roots and return it to the atmosphere in a process called *transpiration*. Water that percolates into the ground can recharge groundwater aquifers. Excess infiltration percolates slowly, moving downward and laterally to sites of groundwater discharge, such as springs on hillsides, or it seeps in the bottoms of streams and lakes or beneath the ocean. If the rate of precipitation exceeds the rate of infiltration, overland flows occur, carrying water into surface water bodies.

SURFACE WATER QUALITY

All the natural and human factors influencing water quality apply to surface water either directly or through surface water and groundwater interconnections. The components of water quality that principally affect surface waters are dissolved oxygen, eutrophication (the reduction of dissolved oxygen in water), temperature, and the presence of toxic substances. These components affect the transmission of communicable diseases through consumption or contact. The dissolved oxygen concentration in a surface supply is an indicator of the general health of the aquatic ecosystem. Toxic substances can affect ecosystems and harm humans who consume contaminated water or contaminated food that comes in contact with the water source. There are increasing concerns over the presence of pharmaceuticals in surface water systems affecting aquatic organisms as a result of human consumption and agricultural runoff. Temperature influences all biological and chemical reactions, and variations in temperature affect the density and transport of water.

Rivers, streams, lakes, and reservoirs are the four principal surface water sources. Estuaries, bays, and harbors are surface waters that connect streams or rivers with the ocean. A river or stream is a surface water source with a distinct downstream flow. A lake or reservoir is a surface water source distinguished by slow flow-through velocities and vertical temperature gradients.

Rivers and streams are rich and diverse ecosystems. The characteristics of the flow and the chemical components of rivers and streams are the principal factors that determine the biological components of water quality. The topography of rivers and streams affects the velocity, mixing characteristics, solids and sediment transport, and flow of the water. All these factors, in addition to the physical, chemical, and biological components, should be considered for their effects on the water quality of a river or stream source.

Lakes and reservoirs are sources for municipal and industrial uses, including recreation, agriculture, water quality control, and fisheries management. Lakes and reservoirs can become sinks for nutrients, toxicants, and other substances from incoming rivers; as a result, eutrophication is a significant water quality problem. The drainage area and length, depth, surface area, and volume of a lake or reservoir are significant factors affecting the extent to which the source will act as a sink.

Wetlands are flooded areas that are shallow enough to promote the growth of wetland plants. They can remove urban pollutants and raise the level of dissolved oxygen by decreasing the levels of oxygen-demanding substances. Contaminant levels can build up in wetland sediments, and some contaminants can be taken up by wetland plant species, concentrating the contaminant while removing it from the water.

GROUNDWATER QUALITY

All the natural and human factors influencing water quality apply to groundwater either directly or through surface water and groundwater interconnections. Groundwater is principally affected by seepage or leaching of minerals and/or contaminants through soils and by pumping. Iron, manganese, high hardness, radon, volatile organic chemicals, inorganics such as nitrate, and pesticides are frequently found in groundwater. While iron, manganese, arsenic, radon, and other minerals *originate* in the geology of the groundwater, the other contaminants are *transported* to the groundwater. High chloride levels have been found near saltwater bodies where fresh groundwater is pumped faster than the water is recharged. If the water table lowers, salt water can intrude. Pumping water with underlying chlorides can cause the high chloride water to upcone into the well. Similarly, high pumpage rates can induce movement of other naturally occurring elements and contaminants into a drinking water well.

Geological and hydrological conditions, as well as the contaminants, make each groundwater source unique. Groundwater flow is very slow laminar flow (compared to surface flow) with no mixing or dispersion of contaminants. The flow may be turbulent near wells and interconnections with surface water. In turbulent regions, contaminants mix and disperse. By the time the water from a contaminated aquifer reaches the well and is recognized, the aquifer has generally been extensively contaminated. The contaminants are usually transported through the soils above the groundwater source from contaminant sources such as agriculture fields, landfills and other hazardous waste sites, and leaking underground storage tanks that contain chemical products. The movement of water through soils depends on oil characteristics such as permeability and moisture content. Other sources of movement include groundwater recharge and injection wells.

The transport of contaminants in groundwater depends on the dilution and mixing characteristics; adsorption, permeability, precipitation, and hydrolysis properties; and the decay characteristics. Hydrocarbons such as methyl-*tert*-butyl-ether (or MTBE, a gasoline oxygenate additive) can move very quickly through groundwater sources.

The US Environmental Protection Agency (USEPA) defines *underground source of drinking water (USDW)* in the Code of Federal Regulations (40 CFR 144.3), which should be considered in the development of groundwater supplies. In addition, guidelines on groundwater monitoring are found in AWWA Manual M21, *Groundwater* (AWWA 2014).

PHYSICAL, CHEMICAL, AND BIOLOGICAL COMPONENTS INFLUENCING WATER QUALITY

Pure water is tasteless, colorless, and odorless. Because water is a nearly universal solvent, organic and inorganic materials, including pollutants and gases, can become components of water as it moves through the hydrologic cycle. The quality of a source is measured by the kinds and quantities of these components found in the water. The components are generally categorized as physical, chemical, or biological.

The USEPA has established a multitude of regulations that protect and regulate the quality of water to be used as a source of water supply. Examples include the Safe Drinking Water Act (SDWA) Stage 2 Disinfectants and Disinfection Byproducts Rule, Groundwater Rule, and the Long Term 2 Enhanced Surface Water Treatment Rule (LT2ESWTR), as well as elements of the Clean Water Act.

Physical Characteristics

The principal physical characteristics of source water include total dissolved solids (TDS), turbidity, color, tastes and odors, and temperature. The following sections provide a brief discussion of each of these parameters.

Total dissolved solids. TDS refers to the total inorganic and organic particulate material in water. TDS is often used as a measure of mineral content and can be estimated with electrical conductivity tests. USEPA 1987 states that "fresh/potable water generally refers to TDS contents of less than 3000 ppm, whereas USDWs include water with <10,000 mg/L total dissolved solids (TDS)." The guidance for USDWs further stresses the need for site-specific information. Although it is not a regulated measure of water quality, the USEPA has established a nonenforceable secondary standard of 500 mg/L.

Turbidity. Turbidity in water results from suspended matter that causes the water to be opaque or cloudy. The suspended matter is often soil runoff in water sources. By itself, turbidity has no health effects, but it can interfere with disinfection and provide a medium for microbial growth. Turbidity usually indicates the presence of microbes. The USEPA Surface Water Treatment Rule (SWTR) requires systems using surface water, or groundwater under the direct influence of surface water, to filter the water when certain water quality parameters are exceeded (see chapter 8). If turbidity goes above 5 nephelometric turbidity units (ntu), the system must filter the water. Systems that filter must ensure that the turbidity goes no higher than 1 ntu (0.5 ntu for conventional or direct filtration) in at least 95 percent of the daily samples of any two consecutive months.

Color. Color in water is caused by natural metallic ions, certain types of dissolved and colloidal organic matter leached from soil or decaying vegetation, and industrial wastes. The USEPA has established an aesthetic secondary standard of 15 color units (with higher color units denoting a more perceptible discoloration). Highly colored water discourages people from drinking it, a problem for water purveyors. Because substances that cause color are usually in solution, the color generally cannot be removed by mechanical filtration.

Tastes and odors. Tastes and odors are caused by salts or TDS, decomposed or synthetic organic material, algae, decaying vegetation, and volatile chemicals. Tastes and odors are usually more closely related with biological properties of water than chemical properties. Often, tastes and odors associated with dead organic matter can be broken down into taste- and odor-free compounds in water with adequate dissolved oxygen, although many are highly resistant to oxidation. Odor-producing chemicals include solvents, pesticides, and benzene compounds.

Temperature. The temperature of water affects biological activity rates, oxygen saturation, and mass transfer coefficients (which describe how molecules of a substance move across an interface from one phase to another). In the case of groundwater, a significant or relatively rapid shift in temperature, which closely correlates with meteorological or surface water conditions, indicates that the source is influenced by surface water and is subject to the SWTR. Water temperature is also an important environmental consideration; certain fish species thrive in a relatively narrow temperature range—trout, for example, like cold water. Anything above 55°F (13°C) is not preferred by trout.

Chemical Characteristics

The principal chemical characteristics of source water include pH, alkalinity and acidity, dissolved cations and anions, hardness, conductivity, and levels of carbon dioxide. Specific elements or chemicals that affect water quality include nitrates and ammonia, phosphate, chloride, sulfates, organic carbon, iron and manganese, pesticides, solvents, and radioactive substances.

pH. pH is a scale measure of the acidic or alkaline (basic) nature of the water, which ranges from 0–14, with 0 being acidic, 14 basic, and 7 neutral. pH is the negative log of the concentration of hydrogen ions, and *pH* refers to the potential of hydrogen. The pH of natural rainfall is about 5.6. Naturally occurring substances, such as carbon dioxide or minerals, and artificially occurring pollution, such as nitrogen oxides and sulfur dioxide from air emissions, can react with water to reduce the pH of rainfall to as low as 3 (i.e., "acid rain").

Closely related to pH are the characteristics of water called alkalinity and acidity. Alkalinity measures water's ability to neutralize acids, that is, its ability to react with hydrogen (H^+) ions. Acidity measures water's ability to neutralize bases or its ability to react with hydroxide (OH^-) ions. For drinking water, USEPA has established a nonenforceable secondary standard for pH of 6.5–8.5.

Cations. Cations are positively charged ions in solution, and anions are negatively charged. Common cations found in water include calcium (Ca^{2+}), magnesium (Mg^{2+}), and iron (Fe^{2+}). Common anions include chloride (Cl^-), bicarbonate (HCO_3^-), and carbonate

(CO_3^{-2}). Cations and anions indicate the level of, or potential for, certain kinds of pollution or harmful substances in water as well as other characteristics such as hardness. A number of similar numerical scales for rating water hardness have been devised and published, including, for example, the following scale (AWWA 2014):

Hardness Range (mg/L of $CaCO_3$)	Description
0–60	Soft
61–120	Moderately hard
121–180	Hard
More than 180	Very hard

Conductivity. Conductivity is a measure of electrical resistance, the property of a substance to conduct (carry) heat or electricity. The unit of measure is the Siemens (formerly called *mho*), which is the reciprocal of resistivity (1 divided by resistivity). Conductivity provides an estimate of the TDS or can be used to verify the TDS results obtained in physical analyses of water.

Carbon dioxide. Carbon dioxide (CO_2), a minor gas in the atmosphere, is an end product of biological decomposition. In solution, it decreases pH (see previous discussion) and is a measure of the corrosiveness of water. High carbon dioxide levels indicate microbial activity.

Biological Characteristics

No source water—ground or surface—is free from pathogens, although the characteristics may be different. High concentrations with variability exist with surface sources as a result of runoff from animal husbandry operations, septic tanks, stormwater runoff, sanitary sewer overflows, and other discharges. Overflows from combined sanitary overflow systems during severe storm events can also introduce pathogens. Pathogens—including bacteria, viruses, and protozoa—can impair water supplies.

Bacteria. Bacteria are unicellular organisms and exist either as free living or parasites. Some bacteria are disease-causing pathogens. In theory, any organism that can invade the human body and produce a toxic reaction in tissues or organs is a pathogen; in practice, to be considered as an *outbreak* of waterborne pathogens, the toxic reaction must occur in two or more individuals and be spread by water.

Major epidemics resulting from contaminated water supplies have been documented from ancient civilizations, through the Middle Ages, and into the 21st century. Water treatment resolved most of these outbreaks. Yet there are new or emerging pathogenic bacteria that are the results of improvements in laboratory detection methods or of a species mutating into a more aggressive invader of humans. The pathogenic bacteria *Escherichia coli* 0157:H7 (*E. coli*), previously considered an emerging pathogen, is an example of the latter group. "Opportunistic pathogens" are bacteria that can harm individuals with weakened immune systems, such as the elderly, individuals with compromised immune systems, and infants with underdeveloped immune barriers.

Viruses. Human viruses found in source waters include rotavirus, enteroviruses (such as polio), reoviruses (which can infect the respiratory and gastrointestinal tracts), hepatitis A, and gastroenteritis viruses (including hepatitis E). All these organisms present significant health concerns and, if present in source water, must be removed or inactivated by water treatment processes. They are productive at lower temperatures but require a host to multiply or move.

Viruses occur more frequently and in greater abundance in surface source waters, but they may also be found in groundwater as a result of septic tank effluents, land application of sewage and sewage sludge, and other sources. Viruses can travel long distances as water percolates downward and flows through the soil; the distance of travel depends on the type of soil and the specific geology. If a groundwater source is directly influenced by surface water intrusions, full conventional treatment may be necessary to ensure the treated water is safe for consumption. The US SDWA established a maximum contaminant level goal (MCLG) of zero for enteric viruses in finished water.

Protozoa and parasites. Waterborne parasites, such as *Giardia* and *Cryptosporidium*, are acknowledged as the leading identifiable agents of diseases acquired from drinking water. *Giardia lamblia* was first observed by van Leeuwenhoek in 1681, but not until the 1960s was *Giardia* recognized as a significant cause of human disease. The SDWA established an MCLG of zero for *G. lamblia* in finished water. *Cryptosporidium parvum* was first described by Tyzzer in 1912 but was not acknowledged as a significant threat to human health until the 1990s.

Algae and fungi. Algae is generally a surface water issue. Algae grow only in the presence of sunlight, and the presence of algae indicates eutrophication. Green algae and cyanobacteria (blue-green algae) impart color, taste, and odor to water. When algae die, the organic matter of the water increases the biochemical oxygen demand—the amount of oxygen taken up by microorganisms in decomposing the organic material in a water sample stored in darkness at 68°F (20°C). Some algae, such as cyanobacteria, can produce toxic substances. Fungi can grow without sunlight, impart taste and odor to water, and may clog water supply appurtenances.

Treatment of affected waters (typically with chemicals) is unreliable. Preventing algae growth by reducing the nutrient loading—especially phosphorus—is the best solution. The problem is worse in summer, when thermal stratification of reservoirs may lead to anaerobic conditions (an absence of free oxygen) in the lower layers. Blue-green algae are tolerant of such conditions. One method to reduce algae growth during the summer is to introduce river water upward by jets into a reservoir. The mixing will reduce the anaerobicity in the lower levels.

SOURCES OF WATER CONTAMINANT LOADING

Each type of water use has a unique set of water quality requirements or criteria. The main categories of water use are water supply (municipal and industrial), recreation (swimming, boating, and aesthetics), fisheries (commercial and sport), agriculture (irrigating fields or watering livestock), and ecological balance. Water used for potable supplies is regulated most stringently; it must be safe, that is, without risk of adverse health effects. Potable water has historically been classified as the "highest and best" use of a water source, although that classification has been questioned because the needs of ecosystems are often much higher quality than that needed for human use. Water used for recreation must be generally safe for ingestion or contact. Water used for fisheries must support aquatic species and be free from contamination that could pose a human health risk. Water used for agriculture should not adversely affect the agricultural use (plants or livestock). Water used to maintain ecological balance, such as minimum stream or river flows, is becoming a more prominent factor when evaluating water availability.

Natural and human factors influence the quality of a water source. Natural factors include climate, watershed characteristics, geology, microbiological growth, fire, saltwater intrusion, and density or thermal stratification. In general, natural factors cannot be controlled. Human factors are categorized as "point" or "nonpoint" sources. Point sources are identifiable localized sources of contaminants and include wastewater treatment plant discharges, industrial discharges, hazardous waste facilities, mine drainage, and spills and releases. There have been increasing concerns over pharmaceuticals entering surface

Table 6-1 Water states and qualitative characteristics

Water State	Nutrient Level	Microbiological Activity
Oligotrophic	Low	Minimal
Mesotrophic	Moderate	Moderate
Eutrophic	High	High

waters through wastewater discharges and agricultural husbandry operations. Effects to the biological communities have been documented. Unregulated contaminant monitoring, proposed by the USEPA in March 2011, aims to document the presence of these as well as other contaminants in drinking waters.

Nonpoint sources come from many diffuse sources, including agricultural and urban runoff, livestock, land development, landfills, erosion, atmospheric deposition (such as acid rain), and recreational activities (such as boating on reservoirs). Because of the larger areas involved and their diffuse nature, nonpoint sources of contamination are more complex and difficult to control than point sources. As point sources become more controlled or treated before release to a water source, nonpoint sources are increasingly being studied and regulated.

Natural Influences

Precipitation can affect water quality, and rain or snow can contain contaminants from air pollution. Storm runoff may dislodge soil particles and pollutants and carry them into surface water, increasing the turbidity, color, metals, or other contaminants. The natural flushing effect of storm flows also introduces particulate organic material and leaches dissolved ions from the contributing drainage area. Dry conditions can cause stagnation, increasing the likelihood of microbial activity and algal growth.

Watershed characteristics such as topography, vegetative cover, and wildlife can significantly affect water quality. Topography affects flow rates and the pattern of runoff and can lead to the introduction of sediment, debris, and nutrients into a water source. Natural decomposition of vegetation can affect water color and is a source of natural organics that contribute to the formation of disinfection by-products. Vegetative cover can act as a natural filtering mechanism for runoff from nonpoint sources and can provide a barrier to human activities. Wildlife can affect the microbiological characteristics of surface waters.

The geology of a watershed directly affects the quality of surface water and groundwater. The weathering characteristics of the geologic parent material affect soil characteristics such as soil depth, texture, and structure—all of which affect natural erosion rates. Soils act as buffering agents for acidic precipitation that can adversely affect biological activity in lakes and reservoirs. Geologic formations in groundwater can release calcium and magnesium minerals, thus increasing the hardness of the water. Uranium may decay, releasing radon. Arsenic and other metals can also be leached into water from certain geologic formations in a watershed.

The state of a water body depends on nutrient levels and microbiological activity. Table 6-1 introduces the three water states and their characteristics.

Point Sources

Twenty-five years ago, sewage treatment plants served only 85 million people in the United States. Today, the number of people served by modern wastewater treatment facilities has more than doubled to more than 180 million people. Wastewater plants have only been regulated since 1946, so while sewage discharges from municipal and community

Source: *Belovodchenko Anton, Shutterstock*

Figure 6-1 Point source pollution

wastewater systems remain major sources of nutrients and organic contaminants, they have improved dramatically in the past 30 years.

Discharges to water bodies with slow-moving, warm waters can become eutrophic as a result of wastewater with high concentrations of nitrogen and phosphorus. Continuing urbanization of the world will further affect the quality of our waters. New constituents, like pharmaceuticals, are being introduced into surface waters through wastewater treatment discharges. Aside from effects to the biota in receiving waters, these compounds can be difficult to remove in the drinking water treatment process.

Industrial facilities can also discharge directly into a water supply either through a regulated discharge or accidentally, as in the case of a spill. Treatment and storage lagoons can contaminate groundwater through subsurface percolation. Figure 6-1 illustrates an example of point source pollution.

Operating a hazardous waste facility in a water supply watershed or aquifer requires extensive precautions to prevent hazardous contamination to the water source. Hazardous substances may be flammable, corrosive, reactive, or toxic. Inactive facilities may require cleanup.

Mine drainage may cause a change in acidity and cause iron, manganese, and other contaminants to suspend in the water; even historical mines that may have been abandoned 150 years ago may pose ongoing risks to water supplies. Mining operations may also affect the topography of a watershed. Watersheds traversed by major highways and rail freight lines are subject to transportation accidents and chemical spills.

Many spills and releases occur in conjunction with chemical product storage installations. Leaks and spills from underground storage tanks and piping are of special concern because the release may go undetected for a long time.

Source: *B. Brown, Shutterstock*

Source: *Nataly Studio, Shutterstock*

Figure 6-2 Nonpoint source pollution

Nonpoint Sources

Agricultural runoff is a major contributor of nutrients, pesticides, herbicides, and turbidity and microbial constituents in many watersheds. Agricultural operations are often upstream of water purveyors. Erosion, which can be exacerbated by improper tilling techniques, can cause increased sediment loading and increased color and turbidity in the receiving waters. Figure 6-2 illustrates two examples of nonpoint source pollution.

The presence of livestock in watersheds or aquifer recharge zones has a direct effect on bacterial contamination. Feedlots can contribute nitrates to wells and pathogens to both wells and, more particularly, surface supplies. Uncontrolled grazing or overgrazing can eliminate vegetative cover.

Other nonpoint sources include urban runoff from highways, streets, and commercial areas that can contribute substantial loads of nitrogen, phosphorus, suspended solids, dissolved solids (salinity), microbial constituents, heavy metals, and organic contaminants to the receiving waters. This type of contamination can be most severe during a "first flush" storm event that follows a prolonged drought during which pollutants accumulate. On-site domestic wastewater management practices, if not designed, operated, and maintained properly, can also be a source of contamination.

Land development on watersheds changes the topography and vegetative cover, and results in new water uses. Eliminating vegetative cover can result in increased erosion. Impervious surfaces, such as pavement, decrease percolation into groundwater and introduce pollution associated with human activity.

Leachate from landfills can change water quality in many ways. If the landfill does not have effective barriers against precipitation or groundwater movement or a leachate collection system, leaching of any contaminants in the landfill can occur. The extent of contamination depends on the characteristics of the landfill.

Airborne contaminants from industrial air emissions, automobiles, or other similar sources can be transported great distances through the atmosphere. These contaminants may pollute water sources through precipitation (e.g., acid rain). Although the extent of airborne contaminants has been increasingly regulated and controlled, sources do remain. The effects on water quality depend on the nature of the contaminant and the physical, chemical, and biological components of the water source. For example, more pristine water bodies having little buffering capacity can be drastically affected by acid rain.

In response to mounting evidence that air pollution contributes to water pollution, Congress amended the Clean Air Act and included provisions that established research and reporting requirements related to the deposition of hazardous air pollutants to the "Great Waters." The water bodies designated by these provisions are the Great Lakes, Lake Champlain, the Chesapeake Bay, and certain other coastal waters (identified by their designation as sites in the National Estuarine Research Reserve System or the National Estuary Program). In 1995, sulfur dioxide emissions from 445 power plants in the eastern United States were reduced by 3 million tons as mandated by Title 4 of the Clean Air Act Amendments. National Atmospheric Deposition Program/National Trends Network data formed the basis for a 1996 report showing that these emissions reductions have reduced the severity of acid rain in the eastern United States.

Recreational activities may affect the biological and chemical components of a water source. Recreational use of proposed water supply reservoirs has become a significant issue with new surface water sources. The biggest issues include swimming and boating (including sizes and types of motors).

WATER QUALITY SAMPLING AND MONITORING

Sampling of source water for quality is crucial for planning, just as monitoring of quality is critical when the source water is being used for supply. The frequency, duration, and intensity of water quality measurements depend on the designated use of the source as mandated by federal and state regulations. Continuous sampling and analytical devices are increasingly being used to monitor raw water quality on a continuous basis, also allowing for detection of certain parameters as indicators of accidental spills or releases. These can be useful tools for water utilities. Appropriate agencies should be contacted for regulations once the use of a water source is defined. Statistics are generally used to quantify the quality of sampling results.

The numbers and types of samples collected and tested simply to meet regulatory requirements are generally considered a minimal effort in determining water quality. Increasing the numbers and types of samples collected and tested beyond what is specified by regulations provides data that may be more representative of water quality. These data help planners predict potential water quality and the treatment processes and maintenance programs to prevent deterioration of water quality.

The timing of taking samples is very important. If the sample collection misses a target event (such as a storm flush), information gained from the program may not reflect actual conditions. The best sampling techniques and laboratory quality control practices will not improve the final results. Thoroughly understanding the water source being sampled is critical. This understanding requires developing an adequate history of water quality changes and identifying indicators that will predict changes.

Surface Water Sampling

A sampling and monitoring program must consider the changing activities and conditions in the watershed. Agricultural, recreational, municipal, industrial, and other activities affect downstream water quality. The watershed should be assessed before a final comprehensive sampling and monitoring program is designed and implemented. One key is to monitor for those contaminants that might actually occur in the watershed; otherwise, money is being wasted. Potential point and nonpoint sources of contamination should be inventoried. If a monitoring program has been established before such an assessment, the results of the data may be more difficult to interpret.

Regulated point source discharges generally will not cause a sudden change in water quality. Most operations meet their limits during routine operations. The assumption that a waste treatment facility operating within its regulatory discharge limits will not be a major source of microbial contaminants appears sound. However, the atypical event may create a problem, such as discharges from combined sewers during periods of heavy precipitation and hydraulically overloaded wastewater treatment plants. Information about point source discharges is usually available through local state primacy agencies.

The weather may be the most unpredictable influence on water quality. Precipitation, snowmelt, wind, drought, and temperature affect water quality. Overland flows of precipitation and snowmelt carry contaminants into receiving streams as nonpoint discharge. Precipitation runoff and snowmelt cause point source discharges from storm sewers, combined sewers, and treatment plants. The direction of the wind may cause a stream to discharge into a bay or a lake, changing course to flow over an intake. Drought conditions result in a reduced dilution factor and may attract wildlife and livestock to the major streams used by public water supplies. The influence of discharge from a waste treatment facility is magnified by a reduced dilution factor during droughts. The effect of climate change and rising sea level will also have an effect on the quantity and quality of freshwater around the world.

A surface water condition that may warrant increased sampling is reservoir turnover. Facilities that take water from surface water impoundments are subject to annual reservoir turnovers. Although site specific, turnover typically occurs once in the fall and sometimes in spring. Turnover is caused by the mixing of a formerly temperature-stratified reservoir. Cold oxygenated water drops to the bottom of the reservoir and replaces warmer water that was anoxic. This turnover can cause rapid changes in water quality.

Understanding the land use may help with developing an appropriate sampling program. An example of a misleading data collection program is routinely seen in northern US inland streams subject to agricultural runoff. In some northern agricultural communities, manure is applied to the fields in December and January when the crops are off the fields and the frozen earth supports heavy farm equipment. Immediately following the application, little or no runoff occurs; however, near the end of February or early March, a combination of snowmelt and early spring rains carries contaminants into receiving streams. Location-specific seasonal effects must be considered when scheduling applications to agricultural areas that might affect surface waters during runoff events.

Groundwater Sampling

The key to developing a sampling plan for a groundwater system is to obtain information about the aquifer and the wells in use. Performing a source water assessment or developing a wellhead protection program provides most of the information required to develop a sampling plan. Information regarding well construction practices is invaluable; a poorly constructed well cannot be expected to produce consistent water quality. Typical problem areas in construction include the wellhead, casing, or grouting. Holes, gaps, or other breaches in the systems allow contaminants access to a well. To determine well construction problems, sampling may be required within the first few minutes or seconds of operating the well. Table 6-2 outlines the analyses that should be performed for groundwater in wells.

Sampling aquifer-related quality requires detailed knowledge of hydrogeological characteristics. For example, it may be necessary to temporarily seal portions of the well with inflatable bladders to isolate one section of the aquifer from another. Alternatively, samples may have to be collected at specified intervals during a pumping test to detect variations in water quality. In general, an increase in selected microbial indicator counts or metallic values indicates aquifer problems. For example, a correlation between an increase in indicator microbial counts with some surface activities, such as precipitation, indicates direct recharge or groundwater under the influence of surface water. An observed increase in coliform counts over time may justify the expansion of the sampling program to include other possible microbial contaminants. Additional guidelines on groundwater monitoring are found in AWWA Manual M21, *Groundwater* (AWWA 2014).

Table 6-2 Summary of the minimum parameters to be analyzed in water quality tests for groundwater

Calcium (Ca)	Fluoride (F^-)
Magnesium (Mg)	Phosphate (PO_4^-) (total)
Sodium (Na)	Silica as silica dioxide (SiO_2^-) (total and dissolved)
Strontium (Sr)	Total dissolved solids (TDS)
Barium (Ba)	Total organic carbon (TOC)
Aluminum (Al) (total and dissolved)	Hydrogen sulfide (H_2S)
Manganese (Mn) (total and dissolved)	Free chlorine (Cl_2)
Iron (Fe) (total, dissolved, and ferrous)	Oxygen (O_2)
Potassium (K)	Carbon dioxide (CO_2)
Bicarbonate (HCO_3^-)	pH
Sulfate (SO_4^-)	Temperature
Chloride (Cl^-)	Turbidity (nephelometric method)
Carbonate (CO_3^-)	Silt Density Index
Nitrate (NO_3^-)	

Source: *AWWA 2003; Bloetscher et al. 2005*

SUMMARY

The quality of water sources is as critical for resource planning as the quantity each may provide. The lower the quality the more expensive the source water will be to treat. A thorough understanding of the complex physical, chemical, and biological characteristics of any source water is important for effective water resources planning.

REFERENCES

AWWA (American Water Works Association). 2014. Manual M21. *Groundwater*. Denver, Colo.: AWWA.

Bloetscher, F., A. Muniz, and G.M. Witt. 2005. *Groundwater Injection: Modeling, Risks, and Regulations*. New York: McGraw-Hill.

Rabalais, Nancy N. 2002. Gulf of Mexico Hypoxia, A.K.A. "The Dead Zone". *Annual Review of Ecology and Systematics*. Vol. 33: 235–263.

USEPA (US Environmental Protection Agency). 1987. *Region 5 Guidance 3: Determination of Underground Sources of Drinking Water*. Washington, D.C.: USEPA. https://www.epa.gov/sites/production/files/2015-09/documents/r5-deepwell-guidance3-determination-underground-sources-drinking-water-19870205.pdf (accessed Aug. 8, 2016).

Chapter 7

Source Water Protection

In 1610, Governor Gage of Virginia proclaimed that for Jamestown, Virginia, "[T]here shall be no man or woman dare to wash nay unclean linen, wash clothes, ... nor rinse or make clean any kettle, pot, or pan or any suchlike vessel within twenty feet [6 m] of the old well or new pump. Nor shall anyone aforesaid, within less than a quarter mile of the fort, dare to do the necessities of nature, since by these unmanly, slothful, and loathsome immodesties, the whole fort may be choked and poisoned." Four hundred years later, source water protection can be considered a logical, prudent, and economical approach to protecting drinking water and public health.

In the 21st century, drinking water sources, whether surface water or groundwater, remain vulnerable to a variety of pollution sources. Many of the stressors to surface water and groundwater sources are associated with various land uses and span every sector of our economy. In today's societal preference of increasing resiliency and sustainability, managing source water with the goal of maintaining and improving source water quality is a logical, economical, and effective approach that can provide many benefits to public health protection, economic development, and the environment.

Source water protection encompasses a range of management practices and is a vital component of the multiple-barrier approach to safeguarding drinking water supplies. The preventive nature of source water protection has proven to be cost-effective by minimizing the treatment needed to remove contaminants, as well as reducing the chances of overloading treatment systems. On the other hand, the difficulties in quantifying benefits (e.g., illness and death avoided, along with other nonhealth benefits) have made source water protection a "hard sell." Therefore, by necessity, source water protection is a collaborative process involving multiple stakeholders and partners who are affected by the protection programs.

Source water protection had long been promoted and used by water systems, local governments, and states as a part of their portfolio to ensure that drinking water for public water systems is of high quality and meets drinking water standards for public health protection. Since the passage of the 1974 Safe Drinking Water Act (SDWA), protection of drinking water supply sources in the United States has been directed by Congress and implemented by the US Environmental Protection Agency (USEPA) under the original act and subsequent amendments of 1986 and 1996. Much of the early efforts under the SDWA on source water protection focused on groundwater sources, and treatment was emphasized for surface water sources. The 1996 amendments recognized source water protection, among other measures, as part of the comprehensive approach to ensure the quality of drinking water by protecting it from source to tap. It should be noted that at the federal level, source water protection is not required by law, apart from the initial source water assessments as mandated by the SDWA Amendments of 1996 for states to complete.

In contrast to the United States, Canada has nonbinding guidelines at the national level for drinking water quality. Drinking water has long been considered the responsibility of the provinces. The drinking water guidelines follow the public health approach of other regulatory policies that are protective in nature and address the multiple-barrier approach endorsed by Hill (2004), of which source water protection is one of the barriers. Similar to the United States, source water protection is voluntary in Canada; however, in response to the *Escherichia coli* outbreak in Walkerton, Ontario, in May 2000, the provincial government of Ontario passed the Clean Water Act in 2006 to protect existing and future source water. For the rest of Canada, although source water protection requirements can be found in various provincial laws and regulations, they are not explicitly imposed on water systems. Nevertheless, useful guidance can be found as part of a collaborative research initiative of the Canadian Water Network and the Water Policy and Governance Group (n.d.) at the University of Waterloo titled *Governance for Source Water Protection in Canada*. The voluntary nature of source water protection in both the United States and Canada has made it highly flexible but also challenging to implement.

This chapter consists of three main sections. The first section presents benefits and challenges of source water protection programs (across both surface water and groundwater sources), along with the key elements for effective source water protection efforts. The second section addresses groundwater source protection. Finally, the third section covers issues associated with surface water source protection and linkages to US Clean Water Act programs (e.g., watershed management and water quality improvement).

SOURCE WATER PROTECTION PROGRAMS

Throughout most of human history, protecting drinking water supply sources was the only viable option for obtaining safe drinking water. In modern times—even with advanced technological treatment methods that many water systems rely on for producing safe drinking water—source water protection programs (SWPPs) are often considered the preferred approach rather than using water treatment technologies to remove contaminants such as anthropogenic compounds (e.g., synthetic organic chemicals) and natural contaminants (e.g., arsenic and pathogens). Based on early experiences from the original SDWA of 1974 and the 1986 amendments, Congress passed the 1996 amendments, which required states to assess the vulnerability of all public water systems to contamination. The 1996 amendments also promoted the notion of source water protection, but stopped short of requiring the development and implementation of SWPPs for all water systems. The basic assumption for developing and implementing SWPPs is that a multiple-barrier approach to protect drinking water will provide high-quality water supplies and enhance safeguarding of public health at the tap.

Multiple layers—including source water protection, treatment, distribution system maintenance, monitoring, and consumer awareness—are components that work in concert to protect the quality and safety of the drinking water delivered by public water systems. Source water assessments (SWAs) form the centerpiece of SDWA's focus on the preventive nature of source water protection. The SDWA Amendments of 1996 authorized funds out of the Drinking Water State Revolving Fund (DWSRF) for states to conduct SWAs. By 2006, most of the SWAs were completed (USEPA 2006). Because states had a limited time frame and limited resources, many of the SWAs relied on easily obtained data and may therefore be lacking sufficient details. Nevertheless, these initial SWAs, along with subsequent updates (if available), provide a valuable tool to help public water systems, along with local stakeholders, develop and implement SWPPs. In 2006, USEPA published a manual to help stakeholders update their initial assessments to address science, technology, and management changes that have developed since the initial assessments and to incorporate the availability of more detail information of their source waters (USEPA 2006).

The four main components of SWAs are to

1. delineate the source water protection area;
2. conduct an inventory of potential and actual contaminant sources in the delineated area;
3. determine the susceptibility of the water system to contamination; and
4. make the assessments available to the public.

In addition to funding the SWAs, the SDWA amendments also authorized a fund to implement state drinking water programs and help states provide revolving loans to assist public water systems to improve drinking water infrastructure. States can use up to 10 percent of their annual DWSRF for SWPPs or provide loans to acquire land or conservation easements. An additional 10 percent of a state's DWSRF allotment can be used to administer or provide technical assistance through SWPPs.

Source Water Protection Assessment Components

Delineating source water protection area. Section 1453(2)(A) of the SDWA requires that the states delineate the boundaries of the source water protection areas (SWPAs) from which one or more public water systems receives its drinking water supply. The delineations are to account for all reasonably available hydrogeologic information, such as water flow, recharge, and discharge, and any other reliable information the state deems necessary to delineate such areas. The USEPA's Source Water Protection Final Guidance document (USEPA 1997c) specifies the following delineation approaches:

- For surface water systems, the SWPA should include the entire watershed area upstream of the intake(s) of a public water system, up to the state border. The states can segment large watersheds into subbasins or buffer zones to provide for more cost-effective contaminant source inventories and susceptibility analyses (USEPA 1997c). Figure 7-1 illustrates a watershed area. Figure 7-2 shows an example of the segmentation of a watershed area for assessments.
- For groundwater systems, delineations should be made using one or more of the following methods: (1) arbitrary radii, (2) calculated fixed radii, (3) simplified variable shapes (Figure 7-3), (4) analytical methods, (5) hydrogeologic mapping (Figure 7-4), and (6) numerical flow or transport models (USEPA 1987, 2006). These methods, to a great extent, were inherited from efforts made on the early wellhead protection program (WHPP) under the SDWA Amendments of 1986. This is discussed further in the following section.
- Public water systems that withdraw water from aquifers with strong hydraulic connections to surface water systems (conjunctive systems, Figure 7-5) may require conjunctive delineations. These delineations should account for the zone

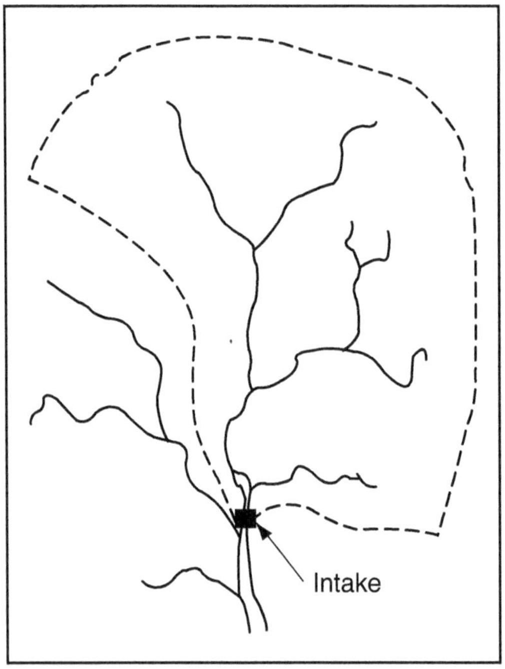

Source: *USEPA 1997c*

Figure 7-1 Watershed source water protection area

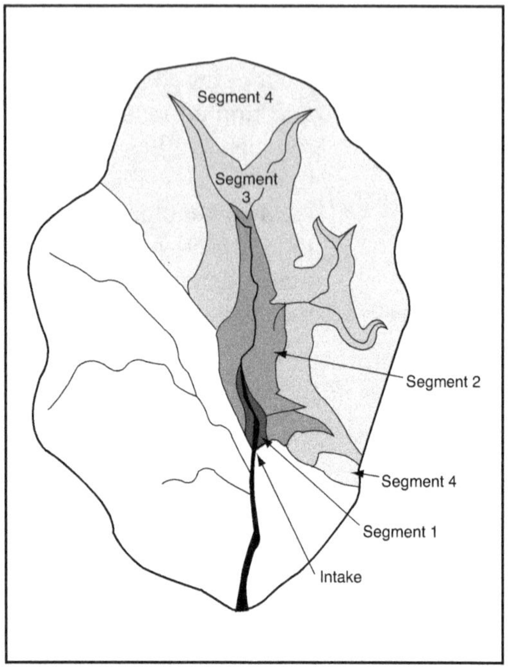

Source: *USEPA 1997c*

Figure 7-2 Watershed source water protection area, segmented for assessments

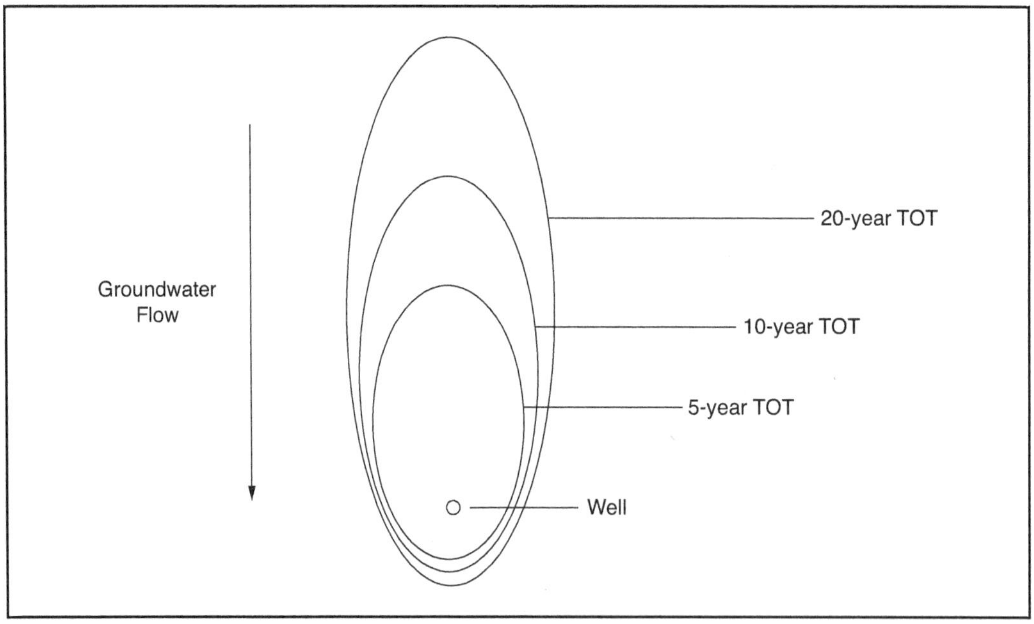

Source: *USEPA 1993*

Figure 7-3 Variable shape source water protection areas with varying time of travel (TOT)

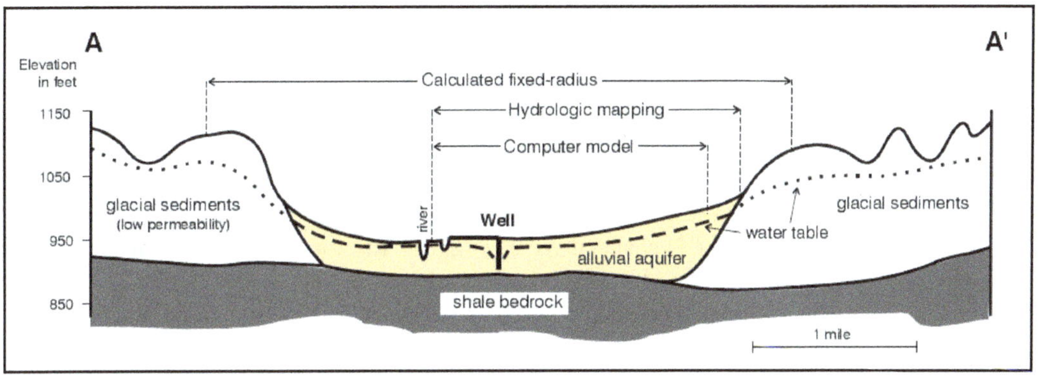

Source: *Keith and Howes, n.d.*

Figure 7-4 Cross section of the alluvial aquifer shows hydrogeologic conditions that affect groundwater movement and the cross-valley extent of each delineation method

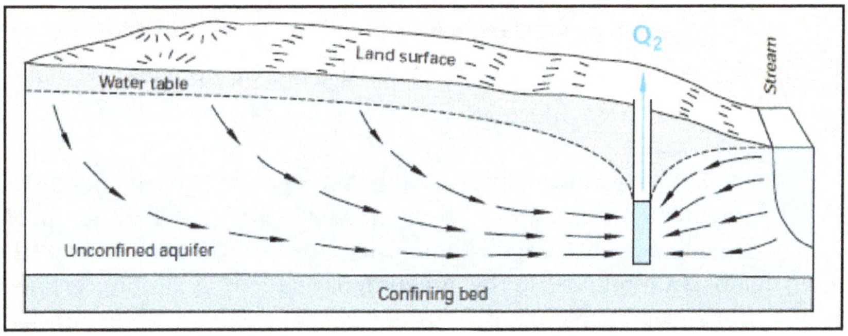

Source: *Winter et al.* 1998

Figure 7-5 Example of surface water and groundwater interaction that would require conjunctive delineation

of groundwater contribution and the area of surface water contributing to the drinking water intake. For example, high-yield wells in alluvial or sand and gravel aquifers often draw water from the groundwater system and from overlying or adjacent rivers or lakes. As a result, the SWPA should include the groundwater system and upgradient surface waters through which contaminants could reach the drinking water intake.

Identifying contaminant sources. States are required to identify the sources of contaminants regulated under the SDWA for which monitoring is required (or any unregulated contaminants selected by the state that may present a threat to public health). These contaminants include those regulated under the SDWA (contaminants with a maximum contaminant level, and contaminants regulated under the various Surface Water Treatment Rules). A variety of regulated and unregulated contaminant sources can be found in SWA areas, including aboveground and underground storage tanks; animal feedlots; agricultural chemical applications and manufacturing; underground injection wells; chemical processing facilities; transportation and road maintenance facilities; septic systems and other onsite wastewater disposal systems; pipelines; and waste transport, storage, and disposal facilities. The range of potential contaminants and contaminant sources is described in several USEPA publications (USEPA 1991, 1993) and are available from numerous state and USEPA websites (e.g., USEPA 2014a, https://www.epa.gov/sourcewaterprotection).

More recent guidance on potential source water contaminants is not available through the USEPA, but information about emerging contaminants can be found in the scientific literature. The most current list of contaminants of concern can be found on the list of contaminants with Primary and Secondary Drinking Water Regulations, the Candidate Contaminant List, and as part of the Unregulated Contaminant Monitoring Program (USEPA 2009, 2014b, 2014c). Because of the wide variety of potential contaminant sources that may be found in SWA areas, a hierarchical approach to source identification is appropriate and most cost-effective. The information sources available to support this approach are summarized here.

Step 1. Search federal and state relational databases to identify regulated sources. The USEPA warehouse allows for online retrieval of environmental information from several USEPA databases:

- Superfund sites—Comprehensive Environmental Response, Compensation, and Liability Information System (CERCLIS)
- Drinking water—Safe Drinking Water Information System (SDWIS)
- Toxic and air releases—Toxics Release Inventory (TRI)
- Hazardous waste—Resource Conservation and Recovery Information System (RCRIS)
- Water discharge permits—Permit Compliance System (PCS)
- Grants information

As a result, an Envirofacts (www.epa.gov/enviro/) and EnviroMapper (www.epa.gov/emefdata/em4ef.home) search allows water systems to rapidly identify existing and abandoned waste management facilities in the SWA area and other ongoing hazardous substance releases to the air, land, or water. In addition, certain states may maintain additional searchable data on underground injection wells, underground storage tanks, oil and gas wells, coal mines, and other potential hazards. USEPA maintains and updates a number of environmental databases. For example, the Superfund site information database (CERCLIS) was migrated to the Superfund Enterprise Management System (SEMS) in 2014 (USEPA 2014d). Leaking underground storage tank databases are maintained by state Departments of Environmental Quality or Natural Resources.

Step 2. Collect information from local land records, sanitary surveys, and/or public health records. Information can be collected from local government records, including operating, discharge and disposal, construction, and other permitting information; zoning records; real estate titles and transactions; and health department records (e.g., septic and onsite wastewater disposal system permits). Maps, aerial photographs, telephone directories, and historic records can be used to locate particular land uses that have been or are currently threatening source water.

Step 3. Collect new information on past land-use practices or contaminant sources that have not been identified. Because many sources, such as product storage facilities, failing septic systems, or abandoned underground storage tanks, may not be tracked at the local level, additional information must often be collected to adequately characterize threats to source water. Such information can be collected by door-to-door canvassing within the SWA area, windshield surveys, mail surveys, or general public education and outreach to request and gather new information. A number of communities have successfully used local volunteers to collect such information. Many communities also regularly compile high-resolution aerial photographs that can be used to identify potential sources of contamination and to update old databases.

Step 4. Target significant sources for further investigation. USEPA (1991) provides worksheets to help target more significant contaminant sources based on the type and volume of materials managed at the sources, and the proximity and vulnerability of the drinking

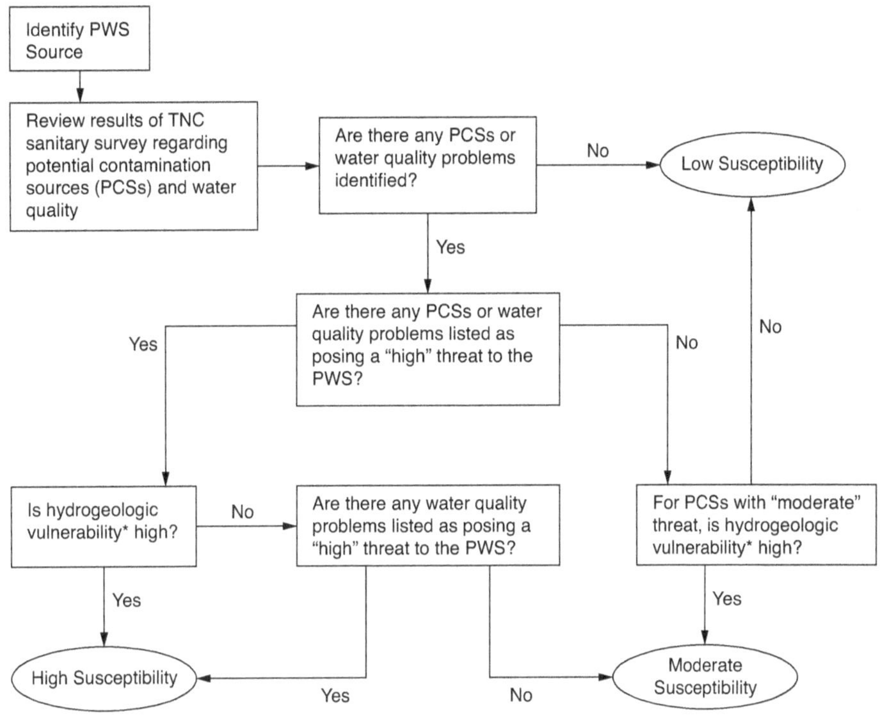

* Hydrologic/Hydrogeologic Vulnerability High = all surface water sources and wells in unconfirmed and semi-confined aquifers; Moderate = wells in confined aquifers.

Note: Susceptibility determination findings will be accompanied by a discussion of the potential risks to the public of drinking water of impaired quality. The risk information breach system will include water quality, treatment, monitoring, and protection measures.

Source: *Massachusetts DEP 1999*

Figure 7-6 Massachusetts Department of Environmental Protection's approach to conducting susceptibility determination

water intake. For sources deemed most significant, the accuracy and reliability of the gathered information should be verified through field checks.

Assessing vulnerability and determining susceptibility. Susceptibility assessments are the least understood (and most important) aspect of source water protection. Assessments are intended to identify the contaminant sources that pose the greatest threat to the drinking water supply so they can be targeted for management. Assessment methodologies include simple, analytical techniques such as hydrogeologic and hydrologic mapping to identify the relative vulnerability of groundwater and surface water supplies, decision tree analysis (Figure 7-6), and complex contaminant transport models linked to risk assessment matrices. Surface water assessments have relied on predictive models such as Enhanced Stream Water Quality Model (QUAL2E), River and Stream Water Quality Model (QUAL2K), and the Storm Water Management Model (SWMM). Packaged assessment products, such as USEPA's Better Assessment Science Integrating Point and Nonpoint Sources (BASINS), provide an integrated approach to incorporate point source and nonpoint source pollution into SWAs. An inventory of hydrologic models, including links to their recommended uses and technical contacts, is maintained by Texas A&M University (TAMU 2014).

Making source water assessments available to the public. Source water protection initiatives require outreach, education, and other nonregulatory tools. Public outreach and education designed to influence behavior have been effective in preventing pollution, conserving water, and developing supplies. Involving the public at the earliest stages of the initiative (e.g., during the SWA phase) builds support for the entire source water initiative. Effective public involvement approaches have been developed as part of WHPPs for groundwater-based systems and for surface-water-based systems.

Public involvement and participation are needed for water systems attempting to implement source water protection and management measures. As the water supply and public utility industries adopt integrated resource planning, and watershed and source water management practices to ensure sustainable, long-term drinking water supplies, the need to build public support for and understanding of critical management issues becomes increasingly important. As discussed in chapter 2, water systems are developing new approaches for communicating with the variety of stakeholders and stakeholder groups that have an interest in water supply development, treatment, distribution, and demand management issues. Furthermore, other water systems and environmental programs are learning how to work with interested stakeholders in meaningful ways. By adopting these approaches, drinking water systems can build stakeholder alliances that will foster better communication with the public and informed water system decision making. The emerging challenges of new water supply development, source water quality management, and the consumer awareness and public outreach provisions of the SDWA make the need for such stakeholder involvement and alliance building especially critical.

Water systems have historically interacted with a variety of stakeholder groups. At the local level, water systems regularly deal with individual customer service and rate issues. Such local stakeholders may also be represented by developers, subdivision managers, or other bulk purchasers of residential water supply. High-demand users, such as industry or large public or private institutions, may also band together to represent the interest of such users. Finally, public utility commissions or citizens' utility boards are often charged with representing local residential water users to determine the adequacy of water supply rate structures and public utility spending plans.

On a regional level, water supply development and management initiatives intersect with the activities of municipal, township, and county governments; planning authorities; homeowners' associations; and agricultural and industrial trade groups. These stakeholder groups may become critically involved in developing and managing water supply watersheds or regional aquifers that encompass multiple political boundaries. Effectively communicating with such a diverse range of interests can drain the time and resources of even large water systems.

Finally, new monitoring and reporting requirements mandated under the SDWA will change the relationship of individual water systems with state and federal regulators and will require new approaches for communicating with the public. In addition, a growing number of water systems are choosing to work with national programs, such as Groundwater Guardian, which focus on community-based water supply protection.

The growing need for interaction with these stakeholders has led water systems to conclude that public awareness of utility concerns is a key to obtaining support for management initiatives and is critical when responding to the need for new resources and system improvements. Customers and other stakeholders are more likely to buy into management decisions when they are consulted early in the decision-making process. As a component of the SWA process, Congress requested that the results of the assessments be made available to the public. Because of sensitive information associated with various aspects of water systems (e.g., intake and well locations), some states decided to share the assessment results on an "as needed" basis after the terrorist attacks on the United States on September 11, 2001.

The SDWA Amendments of 1996 do not require management controls; however, many water suppliers and local governments are actively controlling contamination sources and preventing new contamination sources from threatening drinking water quality through various management controls. Because of the voluntary nature of SWPPs, in 2006, 13 national organizations united to form the Source Water Collaborative (SWC) to protect America's drinking water at the source—in the lakes, rivers, streams, and aquifers tapped for drinking purposes. The goal of the SWC is to combine the strengths and tools of a diverse set of member organizations to act together, and protect drinking water sources for generations to come. As of 2014, the SWC has grown to include 25 national organizations that have been working together to further the goals of protecting sources of drinking water—recognizing that resources are extremely limited, authorities are split, and the actors who can actually protect source waters are diffuse. Individually, SWC members promote the implementation of source water protection as a part of their overall mission. At the same time, each organization recognizes the synergy of coordinated actions and the need for leveraging each other's resources to increase the chances for success over the chances of each entity doing it alone. AWWA is one of the original founding members of the SWC. Valuable information in using a collaborative approach to foster source water protection can be found at www.sourcewatercollaborative.org/.

ANSI/AWWA STANDARD G300 FOR SOURCE WATER PROTECTION

AWWA's Utility Management Standards program provides a means for the user to assess service quality and management efficiency. The standards developed under this program are intended to improve a utility's overall operation and service by identifying appropriate practices, procedures, and behaviors whose implementation will promote effective and efficient utility operations and thus will contribute to protecting public health, public safety, and the environment.

The Utility Management Standards (also known as the G Series) are designed to serve water, wastewater, and reuse utilities and their customers, owners, service providers, and government regulators. The American National Standards Institute (ANSI)/AWWA G300, *Source Water Protection*, published in 2014, provides a framework through which the user can assess its effectiveness at source water protection. The standard was established as a general framework for a utility of any size and circumstance in developing SWPPs that are locally specific and variable in scope. Utilities that fully implement ANSI/AWWA G300 can be assured that they have met the minimum criteria for the initial barrier in the multiple-barrier approach to protecting drinking water. While there may be differences in the details of the programs, ANSI/AWWA G300 applies equally to either groundwater or surface water programs.

ANSI/AWWA G300 lists six essential elements of an SWPP. These six elements describe an adaptive management approach to source water protection (Figure 7-7). In 2010, AWWA published the *Operational Guide to AWWA Standard G300, Source Water Protection* for implementing ANSI/AWWA G300 at the water system level (Sham et al. 2010). The guidebook provided the following expanded description of the six essential elements.

1. **Vision.** A formalized vision guides the development and implementation of an SWPP. A vision statement explicitly expresses the policy of the organization as set forth by the decision-making body of the utility and helps to align priorities and resources. The vision or policy statement is the official declaration of the utility's commitment to source water protection.

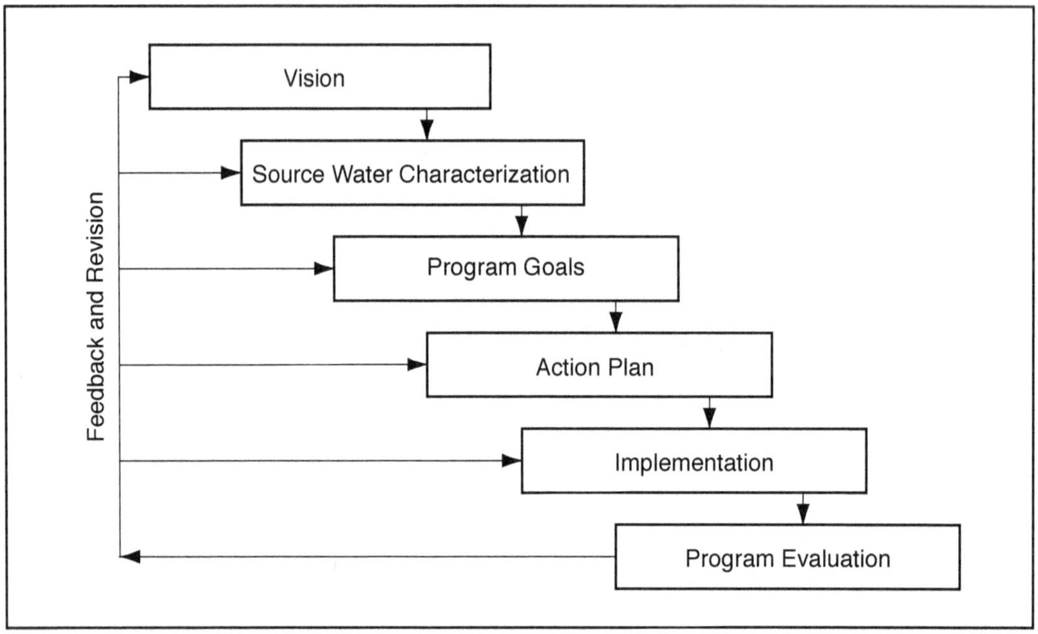

Figure 7-7 Six essential elements of an SWPP under ANSI/AWWA G300

2. **Source water characterization.** Characterizing and assessing both the source water and the land or subsurface area that the source water comes from is essential to obtain the understanding and knowledge needed to develop goals and plans that will realize the source water protection vision. In some cases, it will be appropriate and expected that users will have gone beyond state-performed source water assessments to better define watershed characteristics and will have obtained extensive public participation in defining components of the SWPP. It is also expected that in some cases, other stakeholders may be involved in the characterization process.
3. **Program goals.** Goals and objectives need to be formulated to guide the SWPP and its specific elements. The goals should be targets developed in response to specific problem areas identified through the source water and SWA area characterization and risk assessment processes. The goals should also address each of the drivers motivating the SWPP, including the source water protection vision. Goals may address both current and potential future issues. The goals should be prioritized to reflect the concerns of greatest importance and ideally should specify temporal and qualitative and/or quantitative dimensions (e.g., specific timelines and measurable goals). Both internal and external stakeholders should be involved in goal development.
4. **Action plan.** The action plan lays out a road map of activities to achieve the desired source water protection goals based on the vision, source water area characterization, and susceptibility analysis. The plan identifies required actions (regulations, agreements, practices, etc.) to mitigate existing and future threats to source water quality, and develops priorities and a timetable for implementation, identifying necessary resources and the means for obtaining those resources (e.g., funding), and metrics for measuring the success of each component of the plan. These priorities may be based on the perceived risk from different contaminant sources, the available resources to implement actions, the likelihood of success of different actions, and the obstacles to success that exist for different contaminant sources and action plans.

5. **Implementation.** Implementing the action plan is the core of any SWPP. Planning without implementation does not provide results, and without this step, no actual protection takes place. Developing a comprehensive and implementable plan using an adaptive and iterative management approach to respond to unexpected challenges and barriers and adhering to an established timeline are integral to the success of implementing an SWPP.
6. **Program evaluation.** Administrative programs of any type require periodic (or continuous) evaluation and revision. A good SWPP will include provisions for reviewing and, if necessary, modifying the water system's source water protection vision, characterization, goals, action plan, and implementation elements. This should be done on a periodic basis and also in response to changes in the SWPA and contaminant sources, performance of implemented programs, and so forth. This step of the process is intended to measure the accomplishment or completion of projects, programs, and activities identified in the action plan and to identify obstacles to success and the means to overcome those obstacles. The evaluation and revision facets of the SWPP should be designed with the idea that source water protection is an iterative and interactive process, enabling the SWPP to be a living document, continuously undergoing improvement and updates.

The guidebook for implementing ANSI/AWWA G300 at the water system level (Sham et al. 2010) provides utilities with a detailed discussion of each of the six elements. The guidebook also contains numerous checklists that can be used to assess utilities' source water protection efforts and to develop and enhance their SWPPs.

There are direct benefits to a water utility for implementing an SWPP. USEPA's Long Term 2 Enhanced Surface Water Treatment Rule (LT2ESWTR), promulgated in 2006, includes a provision whereby a water utility may qualify for a 0.5-log disinfection credit as a result of an effective SWPP. Under LT2ESWTR, drinking water utilities that obtain their source water from relatively pristine surface water supplies that are protected by an effective SWPP may apply to the USEPA for a Filtration Avoidance Waiver, which if granted, would release them from the filtration treatment requirements of the SDWA. New York, Boston, Syracuse (New York), Seattle, and Portland (Oregon) are examples of some of the cities that have Filtration Avoidance Waivers.

GROUNDWATER PROTECTION

For most of the United States, groundwater is considered the most reliable and safest source of drinking water. Local governments and the water supply industry have traditionally been responsible for developing, managing, and protecting groundwater supplies in the United States. Today, public water systems (PWSs) that use groundwater sources are subject to the provisions under SDWA. Some states are more stringent than the federal requirements and thus regulate small water systems. Groundwater systems that regularly serve less than 25 individuals, or with less than 15 service connections, are considered private systems and are not subject to federal regulations.

Because of the general availability of groundwater across the country, many PWSs derive their sources from deep, confined aquifers that provide water that typically requires little to no treatment, although, as discussed chapter 4, may also provide little or no recharge. If groundwater treatment is applied, its purpose is typically to remove iron and manganese, for corrosion control or to reduce hardness.

Treatment for contaminants (such as industrial solvents) resulting from land-use activities, such as industry or commercial operations, may be necessary when a public supply well is contaminated. Groundwater contamination and water quality threats from land-use activities present a challenge for systems using groundwater. Groundwater is considered an economical source to develop because wells could be drilled near the areas to be supplied. Water quality threats from encroachment of land-use activities (such as agriculture, and residential, industrial, and commercial development) introduce organic contaminants, metals, nutrients, and biological pathogens to the land surface and subsequently into the groundwater through percolation and infiltration, or directly into the subsurface via underground injection wells and infiltration galleries. Groundwater contamination often is not evident or obvious and is discovered only after a considerable portion of the groundwater resource has become contaminated. Removing chemical contaminants from water is expensive, and the past practice of abandoning and replacing wells is no longer a viable option in areas with limited water resources and growing populations.

Groundwater protection is key to maintaining safe and reliable groundwater supplies. This section presents the regulatory framework for groundwater protection and outlines a stepwise process for groundwater source protection.

Under the 1974 SDWA, two programs were put in place to help protect US groundwater resources: (1) the Sole Source Aquifer Protection Program, and (2) the Underground Injection Control Program.

The Sole Source Aquifer Protection Program is authorized by Section 1424(e) of the SDWA of 1974. It states the following:

> *If the Administrator determines, on his own initiative or upon petition, that an area has an aquifer which is the sole or principal drinking water source for the area and which, if contaminated, would create a significant hazard to public health, he shall publish notice of that determination in the Federal Register. After the publication of any such notice, no commitment for federal financial assistance (through a grant, contract, loan guarantee, or otherwise) may be entered into for any project which the Administrator determines may contaminate such aquifer through a recharge zone so as to create a significant hazard to public health, but a commitment for federal assistance may, if authorized under another provision of law, be entered into to plan or design the project to assure that it will not so contaminate the aquifer.*

The Underground Injection Control (UIC) Program is authorized by Section 1421 of the SDWA of 1974. Part of the SDWA required USEPA to report back to Congress on waste disposal practices and develop minimum federal requirements for injection practices that protect public health by preventing injection wells from contaminating underground sources of drinking water (USDWs). A USDW is an aquifer, or a part of an aquifer, that is currently used as a drinking water source or may be needed as a drinking water source in the future. Specifically, a USDW supplies any PWS or contains a sufficient quantity of groundwater to supply a PWS, and currently supplies drinking water for human consumption, or contains fewer than 10,000 mg/L total dissolved solids (TDS) and is not an exempted aquifer. It should be noted that unless diesel fuel is used in the fracturing fluid, hydraulic fracturing is exempted from UIC regulations. The effects of unconventional oil and gas production using hydraulic fracturing and directional drilling on groundwater are being studied in shale gas production areas. Sole source aquifer protection and UIC programs are used at various levels to protect the USDWs in the United States. These programs should be integrated into any groundwater-based SWPPs, as appropriate.

Comprehensive State Groundwater Protection Programs

In 1992, the USEPA published guidance for comprehensive state groundwater protection programs (CSGWPP) (USEPA 1992). The goal of this approach is to achieve a more efficient, coherent, and comprehensive approach to protecting the nation's groundwater resources by coordinating operations of all relevant federal, state, and local groundwater programs in a state. Under CSGWPP, USEPA asks states to coordinate their groundwater protection actions through six strategic activities:

1. Establish a groundwater protection goal.
2. Establish priorities to guide program efforts.
3. Define authorities, roles, responsibilities, resources, and coordinating mechanisms.
4. Implement efforts to accomplish groundwater protection goals.
5. Coordinate information collection and management.
6. Improve public education and participation in all aspects of groundwater protection.

CSGWPPs are designed to focus pollution source control programs on preventing contamination of high-priority groundwater, facilitate coordination among the many intrastate programs that protect groundwater, and build a comprehensive approach to protect groundwater that includes all stakeholders. In addition, CSGWPPs strengthen state watershed approaches by providing an essential link between the state's groundwater and surface water protection programs. In some geographical settings, however, these plans may be too broad and there is a lack of resources to implement them. Thus, many PWSs that rely on groundwater for their source water maintain an active approach in maintaining and implementing their own SWPPs as discussed below.

Wellhead Protection Program. In 1986, amendments to the SDWA increased the authority and responsibility for drinking water protection. Section 1428 of the SDWA created the wellhead protection program (WHPP). Subsequent amendments in 1996 further strengthened drinking water protection by establishing the source water protection program (SWPP). Because of the duplicative nature of WHPP and SWPP, many states have replaced their WHPPs with the updated SWPPs or incorporated their WHPPs into their SWPPs. One of the subtle differences between WHPPs and SWPPs is associated with the extent of the aquifer under consideration and is discussed further below. The following paragraphs describe the components and operation of WHPPs and the corresponding SWPPs.

WHPPs emphasize pollution prevention rather than treating drinking water contamination and taking remedial actions once groundwater is contaminated. Each state has developed a program tailored to the specific physical, cultural, and administrative conditions of the area to protect its groundwater supplies. Variability in hydrology, geology, land use, and management approaches have resulted in a wide range of programs to protect public water supply wells. For a state to have an approved WHPP, it must meet the following requirements:

1. Specify the roles and duties of state agencies, local government offices, and public water suppliers regarding development and implementation of the program. Many states have worked cooperatively with public water suppliers and local governments to form local teams to plan and implement wellhead protection efforts. A comprehensive approach, including stakeholders and the public, is recommended to achieve maximum results. A typical planning team could include representatives from local public interest groups (including environmental groups, community service organizations, and public representatives), regulatory organizations (such as health, planning, and public works), public service organizations (water utility, fire department, and extension service), and private organizations (including businesses, farmers, and developers). In some instances, larger water utilities are fully responsible for implementing and periodically revising a WHPP.

Stakeholder involvement is garnered through the public meeting process and sometimes by maintaining a citizen's advisory board.

2. Delineate a wellhead protection area (WHPA) for each wellhead based on hydrogeologic and other relevant information to define the boundaries of the most critical land areas that provide recharge to the well. The techniques used to delineate WHPAs depend on the well type, hydrogeology, management approach, and budget. Areas that receive recharge that contribute water to a pumping well are referred to as zones of contribution (ZOCs). These ZOCs are affected by recharge rates, pumping rates, and hydrogeologic boundaries.

   The methods used to delineate WHPAs include arbitrary fixed radius, calculated fixed radius, simplified variable shapes, analytical methods, hydrogeologic mapping, and numerical modeling. If stringent land-use controls are to be developed and enforced in WHPAs, more detailed and scientifically accurate methods should be used to delineate the area. If the management approach is only to educate the public and landowners of the connection between their land-use activities and water quality in the well, the simple arbitrary fixed radius or analytical methods may be sufficient. Details and examples of delineation methods can be found in several USEPA publications (USEPA 1987, 1988, 1997a, 1997b). An inventory of hydrologic models and technical contact information is maintained by Texas A&M University (TAMU 2014). Individual states have issued more recent guidelines for WHPA delineation including Minnesota (MDH 2009) and Wisconsin (WDNR 2012), two states that, on average, rely heavily on groundwater sources. As one can note, these methods were passed on for use in SWPPs under the 1996 SDWA Amendments. Advantages and disadvantages of various delineation methods are summarized in Table 7-1.

   In the current WHPPs for one midwestern utility, many ZOCs are considered simultaneously and include fixed radii of 1,200 ft (366 m) and ½ mile (4,250 km) and hydrologically modeled time of travel radii for 5, 50, and 100 years. Activities within each of these radii are managed through zoning and cooperative agreements with decreasing stringency as the activity moves further from the well.

3. Identify sources of contamination in each WHPA. The delineated WHPA becomes the boundary within which a land-use inventory can be conducted to identify and map potential contamination sources. While not considered as direct sources of contaminants, abandoned wells and drainage wells should be included in ZOC delineation because they may act as direct conduits for pollutants from the surface to enter an aquifer.

   Not only should current land uses be mapped and assessed for their potential threat to water quality but also for areas of future intended uses on the land. Distance from the well, contaminant type, and contaminant severity are important considerations when assessing water quality threats.

   For example, a few septic systems or onsite wastewater disposal systems in a WHPA may present very little threat to water quality. However, hundreds of onsite systems will definitely create a threat that will most likely result in lowering water quality. Many governmental sources are available to provide data on land-use threats. States and USEPA regional offices have extensive databases providing information on large permitted facilities, such as wastewater treatment plants, landfills, waste sites, and drainage wells. Local and state agencies may also maintain records of gas stations with underground storage tanks.

   Most water quality threats can be identified only by conducting field surveys and inventories. Small commercial operations that handle small quantities of hazardous materials, private septic and onsite wastewater disposal systems, roadway drainage, and underground petroleum storage tanks are all activities

Table 7-1 Advantages and disadvantages of WHPA delineation methods

Arbitrary fixed radius	Little data necessary
	Quick and easy to draw
	Very low cost
	Not very accurate
	Not suitable for confined aquifers
Calculated fixed radius	Need limited hydrogeologic data
	Relatively quick and easy
	Inexpensive
	Not highly accurate
	Not suitable for confined aquifers
Variable shapes	Based on relatively little field data
	Still fairly quick and easy
	If data are available, low cost
	In complex settings, not very precise
WHPA code (semi-analytic model)	Based on substantial field data
	May require technical assistance
	Automatic delineation of capture zones
	Calculates the effects of well interference
	Danger of hidden errors because the program is simple
	Most solutions assume homogeneous isotropic aquifers
	Moderate costs
Analytic models	Based on substantial field data
	Probably require professional help
	Moderate costs, if data are available
	Widely used, fairly accurate
	Suitable for rock wells (if data are available)
Numerical models	Based on extensive field data
	Requires computer/technical expertise
	Can be highly accurate
	Can be quite expensive

that pose a water quality threat but will not be easily found in any centralized database. Examples of contamination threats and inventory approaches can be found in USEPA (1989, 1993) and New England Interstate Water Pollution Control Commission (NEIWPCC 1996, 2004). Further resources and discussion of source characterization activities can be found in the Surface Water Source Protection section later in this chapter. While focused on surface water resources, many of the same principles apply to groundwater.

4. Develop management approaches. Management approaches to protect the water quality of public water supply wells are either nonregulatory or regulatory controls, as well as nonstructural or structural in nature. Nonstructural controls include education programs or zoning prohibitions. Structural controls include physical changes to the natural or built environment, such as the requirement for double liners for underground storage tanks. Nonregulatory approaches include

voluntary changes in behavior through public education, land acquisition, purchasing conservation easements, spill response contingency planning, and increased monitoring. These controls can be extremely effective but may require more creativity and innovative approaches to implement.

Regulatory controls are used to restrict current and future land uses through the use of zoning, subdivision regulations, local ordinance, and health authorities. Many public water suppliers have worked cooperatively with local governments to help enact overlay-zoning ordinances to implement controls in the delineated WHPA. Most of these regulations are enacted to prohibit the siting of extremely hazardous activities in the WHPA or to place conditions on uses that will reduce their potential threats. For instance, industrial activities using hazardous materials may be allowed in a particular WHPA, but the operator of the facility must include design features to contain, control, and monitor any potential accidental releases. The stringency of the control measures is a local issue requiring the involvement of the public water supplier and stakeholders.

For example, in a midwestern city, a century-old industry was surrounded by development as the city grew over the 100 years. The facility was identified to be between the 5- and 50-year time of travel defined by hydrologic modeling to the nearest utility well. The potential for contamination of the aquifer that serves the utility well with industrial chemicals was identified in the vulnerability evaluation. Therefore, by negotiated agreement involving the industry, the utility, and the state and local regulators, with input from neighborhood members, monitoring wells were installed at the property boundary and monitored at regular intervals. Additional guidance on developing action and implementation plans is discussed in the next section.

5. Prepare contingency plans for alternative sources of drinking water supply in the event of emergencies. Alternative water supply planning ensures that drinking water service can continue uninterrupted if contaminants are released into the WHPA. In addition, detailed contingency plans will identify individuals who should respond and the appropriate level of control to minimize and mitigate water quality threats. A series of questions that can provide assistance with developing these plans can be found in the Worksheets section of the *Operational Guide to AWWA Standard G300, Source Water Protection* (Sham et al. 2010).

   Within the emergency preparedness programs, a number of states have developed Water/Wastewater Agency Response Networks that are voluntary. The participating utilities and agencies provide aid to each other in response to natural disasters (such as tornadoes and floods) and significant accidents that disrupt the utility's ability to provide safe water. This type of network has been successfully activated during flooding events along the upper Mississippi River and after hurricanes in Florida.

   Another approach to contingency planning is in Wisconsin, where the state regulators have an agreement in place with companies that operate tankers that deliver milk from dairies to bottling facilities. In the event of an emergency, some of these tankers can be used to provide safe water when a PWS is compromised and is being remediated. In other states, tankers that haul food-quality products could be employed in similar agreements.

6. Identify sites for new wells that will provide protection from contamination. Water supply planning is an important element of WHP because it requires the local water supplier and the community to look ahead and ensure that sites for new wells are protected from potentially contaminating activities. One particularly

important tool used in this planning is a build-out analysis to project maximum growth and water demand, and to identify land-use conflicts.

7. Ensure public participation. Public education is an important element in WHP. Public support, assistance, and involvement will ensure that WHP efforts are successful. Involving and informing the public of WHP activities will help the public understand the need for WHP and provide support to develop and implement the program. If regulatory controls are proposed as a management tool, conducting an effective public education effort is extremely important.
8. The WHPP provides an opportunity for water suppliers, and local and state governments to work cooperatively to delineate WHPAs, identify and assess current and potential contaminant sources, and develop management approaches to provide protection from contamination. Thousands of PWSs across the United States have active WHPPs working cooperatively with local governments to ensure that recharge areas to public water supply wells are protected from sources of potential contamination. A series of questions that can provide assistance with developing public and stakeholder relations can be found in the Worksheets section of the *Operational Guide to AWWA Standard G300, Source Water Protection* (Sham et al. 2010).

A number of principles and strategies for identifying and building stakeholder groups are discussed in the next section. The strategies apply whether the source is groundwater or surface water.

Notably, WHPAs can encompass land areas with different criteria. For example, many states specified areas of specific radii from the wellhead location to ensure that run-on from surrounding areas would not introduce contaminants that can reach the groundwater via the well bores and poorly constructed wells. In other cases, WHPAs are recharge zones or ZOCs to the underlying aquifers that supply water to the pumping wells. Depending on the hydrogeological settings, a recharge zone or ZOC for a deep drinking water well may not be contiguous to the wellhead or may be located a great distance from the withdraw point. A recent US Geological Survey publication, *Factors Affecting Public-Supply-Well Vulnerability to Contamination: Understanding Observed Water Quality and Anticipating Future Water Quality* (Eberts, Thomas, and Jagucki 2013) is an excellent source of information for gaining a better understanding of protecting groundwater resources.

To a great extent, the successes of WHPPs have led to the development of the SWPPs, which include surface water sources. Many of the steps involved in developing and implementing a WHPP are included in SWPPs. Components to be considered in protecting source water for a surface-water-based system are presented in the following section.

SURFACE WATER SOURCE PROTECTION

Surface water bodies such as rivers, streams, lakes, and reservoirs provide a significant amount of water to drinking water systems, especially in areas where there are large quantities of surface water. Under the 1972 Clean Water Act (CWA) and subsequent amendments, the USEPA was given the authority to set effluent limits on an industry-wide (technology-based) basis and on a water quality basis that ensure protection of the receiving water. The CWA requires anyone who wants to discharge pollutants to first obtain a National Pollutant Discharge Elimination System permit, or else that discharge will be considered illegal. In addition, under the CWA, states and Native American tribes are required to specify appropriate water uses that take into account the usage and value of PWS, protection of fish, wildlife, recreational waters, agriculture, industrial use, and navigation. The suitability of a water body is examined by states and tribes for uses based on physical, chemical, and biological characteristics of the water bodies. States and Native American tribes also examine geographical settings, scenic qualities, and economic considerations to determine designated uses for water bodies. If water quality assessments indicate designated uses to be less than those presently attained, states or tribes are required to revise water quality standards to reflect the uses actually being attained. For any water body with designated uses that do not include a "fishable/swimmable" target use that is identified in Section 101(a)(2) of the CWA, a Use Attainability Analysis must be conducted. Every three years, such bodies of water must be reexamined to verify if new information is available that demands a revision of the standard. If new information is available that specifies "fishable/swimmable" uses can be attained, then the use must be designated.

Although "public water supply" is an appropriate use of a water body under the CWA, specific criteria were never developed in the context of water quality standards and designated uses. Therefore, watershed management becomes necessary to reduce or eliminate existing contamination sources and to prevent the introduction of new contamination sources to the source water.

In contrast to groundwater, surface water resources provide service to a much larger number of users that include

- drinking water systems and customers,
- industrial and commercial establishments,
- recreation,
- agriculture,
- transportation,
- logging and mining,
- wastewater treatment and disposal, and
- wildlife habitats.

Often these uses are in conflict with each other. While an SWPP is specific to the protection of sources for PWSs, the needs of the many interested parties cannot be ignored. Therefore, a watershed protection plan (WPP) that includes several practices, policies, ordinances, and bylaws integrated to best meet the needs of several interest groups or users in the watershed (i.e., stakeholders) is frequently needed.

Protecting the water source and meeting the needs of multiple users involves a wide variety of management approaches, including some of the following:

- Regulations, bylaws, and ordinances
- Enforcement actions
- Land acquisitions
- Monitoring
- Emergency response plans
- Wastewater discharge permits
- Herbicide-pesticide and fertilizer application requirements
- Stormwater management
- Road-salting restrictions
- Overlay protection zoning
- Public education
- Recreation access restrictions

- Dedicated funding
- Incentive-based programs
- Stream and lake restoration

Many watershed protection organizations are involved in implementing WPPs, and partnerships must be formed to ensure the success of these plans. From a hydrologic or scientific point of view, the watershed (as defined by its topographical delineation boundary) is the logical unit of water resource management. However, the water utility, the municipal government, and various regulatory and public agencies are not necessarily bounded geographically by the watershed boundary. Watershed and jurisdictional boundaries that do not coincide present a significant obstacle to the watershed protection approach. Many compromises and trade-offs are needed to obtain the cooperative involvement of local, regional, state, and federal entities, both public and private, with source water protection as a priority focus. Overcoming these obstacles is paramount for a successful protection plan, especially in home rule states where land-use controls generally reside with the local communities.

A WPP should be carried out in an adaptive process. Watershed conditions and societal values continually change and protection measures should adapt to the current scientific, technological, and socioeconomic conditions. The iterative process involves monitoring and assessing, selecting management and protection practices, implementing, evaluating, and adapting. Nevertheless, a base plan is always needed as a starting point. The steps in developing a base WPP are shown schematically in Figure 7-8.

Delineate and Characterize the Watershed

Many methods are available to delineate watershed catchments. Examples of watershed SWPAs were presented in Figures 7-1 and 7-2. The methods range in complexity and cost of implementation. Therefore, the water systems and the appropriate regulatory agencies must choose a method based on regulatory requirements, available resources, local hydrologic conditions, and specific goals.

Some states have delineated all their major watersheds within the state and may have basin maps in digital format on a geographical information system (GIS). In addition, the US Geological Survey (USGS) and the Natural Resources Conservation Service have developed a nested system of hydrologic unit areas, which are essentially watershed delineations. GIS data sets of hydrologic unit areas are available for most states. The national hydrography data set, available from the USGS (http://nhd.usgs.gov/wbd.html), includes the hydrologic unit area delineations, as well as all of the streams within the areas. Hydrologic unit areas are fairly large watersheds. Often, utilities or local stakeholder groups are interested in only a portion of a greater hydrologic unit area. Topographic quad maps from the USGS, usually at the 1:24,000 scale, typically provide adequate information to delineate the watersheds. A base map is often made by adding town boundaries, major roads, and other major landmarks to the topographic map. Several thematic maps and spatial data layers may be overlaid on the base map for developing WPPs.

A typical scenario of map overlays might include the following:

- Watershed base map
- Topography
- Hydrography
- Sub-watersheds
- PWS withdrawal points

Delineate and characterize the watershed
↓
Identify players and form a planning team
↓
Develop and implement an educational outreach program
↓
Identify regulatory and community goals
↓
Classify and rank goals
↓
Develop criteria to measure achievement of goals
↓
Identify activities posing a threat to goals
↓
Quantify activities and impacts on goals
↓
Rank threats
↓
Identify control measures to reduce or eliminate threats
↓
Quantify the efficiency and effectiveness of control measures
↓
Identify administrators of control measures
↓
Rank control measures
↓
Develop a watershed/reservoir protection plan
↓
Revise/modify
↓
Implement the plan
↓
Monitor/analyze

Figure 7-8 Watershed protection plan development and implementation flowchart

- Soils susceptible to erosion (developed and maintained by the US Department of Agriculture, Natural Resources Conservation Service)
- Permitted wastewater discharge points, landfills, and hazardous waste facilities
- Local zoning and overlay districts
- Current land use
- Mining operations
- Oil and gas development
- Forestry
- Residential areas
- Commercial areas
- Industrial areas

- Other hazardous or contaminated sites
- Water quality monitoring stations
- Open space (e.g., parks)

Many of these databases are available in a digital format that allows data to be integrated to produce various maps. A GIS also allows spatial analyses to access the proximity of contamination sources and drinking water intake points, which will help determine the susceptibility of water systems to contamination.

Identify Stakeholders and Form a Planning Team

The success of any WPP depends on including the relevant partners and stakeholders affected by implementing the protection plan. Some stakeholders will have much more interest in the planning process, but all relevant entities and individuals should be represented to ensure the planning process is open and transparent. Once the planning team has been formed, subgroups responsible for specific tasks could be formed to leverage available expertise and make the planning process manageable. These subgroups should be formed based on interest and knowledge, as well as on authority and experience. Potential planning team members might include the following:

- Municipal officials—representatives from the health department, fire department, historic commission, emergency responder, conservation commission, zoning and planning commission, board of selectmen or city councils, water utility staff, code enforcement officer, local health department, and town engineer
- Public organizations—watershed advisory groups, lake associations, League of Women Voters, community service groups, trade associations, and local conservancies
- Environmental and planning organizations—Natural Resources Conservation Service, local Soil and Water Conservation Districts, Cooperative Extension Service, USGS, regional and local planning agencies, and state environmental agencies
- Local citizens—farmers, developers, residents, business and industry representatives, and teachers

All parties with an interest in the watershed should be included on the team. Those who are not part of the utility service district but who reside or own businesses or property in the watershed should also be included. The plan should include benefits to these parties and to the utility's customers. Continual communication about the process is critical for success.

Develop and Implement an Educational Outreach Program

An ongoing outreach program is one of the first steps in developing a WPP. The initial objective of the program is to inform people that a WPP is being developed, identify its initial goals, assess how it might affect people in the watershed, and invite interested parties to join the team and contribute ideas.

The second objective of the educational outreach program is to educate the public about the potential effects that everyday activities may have on water quality. When people become aware of those effects, they become partners in implementing the plan. Many homeowners, for instance, do not understand the link between fertilizing their lawns and the eutrophication of surrounding water bodies. Providing this education is only part of the overall outreach program; teaching the public about available control measures or management practices is another part. The user must be able to implement the control measures, which should be cost-effective and specific to the threat. Effective outreach programs keep the public engaged and informed of the status of the ongoing protection plan, and the benefits or payback.

Identify Regulatory and Community Goals

The SDWA and its 1986 and 1996 amendments establish the regulatory requirements most critical to protecting drinking water sources. The 1986 amendments established the WHPP and the 1996 amendments addressed source water protection requiring states to implement USEPA-approved programs to assess threats. It also encourages the formation of voluntary local partnerships to prevent pollution.

A water utility's goal may be to maintain, or even enhance, source water quality. The community may have other goals, such as preserving open space, improving fisheries, increasing recreational opportunities, minimizing land-use regulation, or expanding the tax base. Agreeing on, and establishing, these goals will dictate the strategy for developing and implementing the SWPP.

Classify and Rank Goals

Once the primary and secondary goals have been determined, they must be ranked. The goals may be classified as primary (i.e., to maintain or enhance the source water quality) or secondary (e.g., developing a nonresidential tax base in the community, developing public recreation areas, or minimizing restrictions on land-use choices in the watershed). Because many goals may conflict, the planning team needs to evaluate trade-offs and set priorities. Later in the development of the plan, the identified control measures and their effects on achieving the identified goals will need to be ranked.

Develop Criteria to Measure Achievement of Goals

The criteria chosen to measure achievement of the goals established in the WPP are extremely important and will serve as the benchmark for the plan. These criteria will also dictate the monitoring program to be followed (i.e., the parameters to be measured, sampling locations, required precision, frequency and duration, and program cost). Although knowing the goals of the program is important to establish a sampling program, the utility may need to know the water quality before establishing these goals. In other words, baseline sampling should precede goal setting to ensure that the collected data will help establish goals and ensure that protective actions will meet these goals.

If the primary goal is to maintain the quality of the source water, then the criteria may be the current drinking water regulatory constituents sampled in tributaries to the water supply source; at inlet locations to the reservoir, lake, or stream; at the water treatment plant's intake; and at various locations in the distribution system. The list of constituents sampled will vary from sampling point to sampling point but might include the following:

- Secchi depth
- Dissolved oxygen
- Temperature
- pH
- Total coliform

- Fecal coliform
- Conductivity
- Alkalinity
- Turbidity
- Color
- Algae (identification and count)
- Nutrient series
- Chlorides
- Hardness
- Metals
- Organic series
- Silica
- Microbiology (bacteria, virus, and parasite)

If the goal is not merely to maintain the source water quality but to enhance it (e.g., possibly anticipating more stringent regulations), the sampling program may be expanded to include contaminants of emerging concern. In addition, fecal source tracking can be used to identify sources contributing contaminants from different land-use activities such as animal husbandry, septic systems and other onsite wastewater disposal systems, storm water, and more. The cost of the monitoring program will need to be taken into consideration. Other secondary goals for monitoring criteria would include

- sustaining multiple uses of the watershed (e.g., recreation, timbering, wildlife protection, and fishery enhancement);
- minimizing social, financial, and environmental effects associated with implementing the plan; and
- establishing partnerships of watershed users, each able to participate in decision making.

The criteria may be categorized as general (directly related to the primary or secondary goals) or specific (related to reducing or eliminating a threat, such as putting septic and onsite wastewater disposal system users on a centralized sewerage system or a regional maintenance plan).

Identify Activities Posing a Threat to Goals

Activities that pose threats to the source water in a watershed can be identified, later quantified, and finally ranked by considering several characteristics of the threats:

- Current status of the threat (e.g., immediate or potential)
- Type of contaminants associated with the activity (e.g., microbial contaminants from a failing septic system or sediment from a construction site—two very different types of contaminants)
- Location of the activity (e.g., proximity of the activity to the water supply source or distribution system intake—locations of high risk)
- Magnitude of the contaminant source (e.g., volume of discharge)
- Frequency of the activity (e.g., continuous or sporadic)
- Duration of the activity (e.g., year-round or seasonal)

Table 7-2 Categories of water quality threats

Human-Induced Nonpoint and Point Sources	Natural Sources
Nonpoint	
Septic and other onsite wastewater disposal systems	Erosion
Recreational activities	Animal populations
Stormwater runoff	
Urban runoff	
Logging	
Herbicides, pesticides, fertilizers	
Animal grazing	
Construction	
Gravel mining	
Road salting	
Future development	
Unpaved roads	
Manure application sites	
Point	
Uncontrolled releases—transportation	
Uncontrolled releases—fixed sites	
Barnyards/feedlots	
Gasoline/petroleum storage	
Permitted activities—wastewater treatment plant discharges	
Permitted activities—solid waste landfills	
Permitted activities—groundwater discharges	
Permitted activities—stormwater	
Permitted activities—construction	
Oil wells not properly abandoned	
Other unauthorized activities	

Threats can also be categorized as either naturally occurring or human induced and as point or nonpoint sources.

The procedure for identifying threats should include the following:

1. Review current literature on the nature of various threats to water quality.
2. Review watershed data and available mapping.
3. Interview watershed residents and local, regional, state, and federal regulatory personnel.
4. Review historical water quality data for the watershed.
5. Conduct field reconnaissance of the watershed (sanitary surveys).
6. Review databases for regulated activities (e.g., National Pollutant Discharge Elimination System releases, confined feeding operations, land application of sludge).

A list of typical threats in various categories is shown in Table 7-2.

Quantify Activities and Effects on Goals

To rank threats to the water supply, the threats must first be quantified. The method should be selected based on the relative effect of the threat on the source water supply. One such method was developed to evaluate the Wachusett Reservoir watershed in Massachusetts. The method evaluated each threat by watershed protection zone and region, as well as whether the specific threat was immediate or potential. The watershed was delineated into three zones:

1. Zone A: Areas located within 400 ft (122 m) of the 100-year floodplain elevation as delineated on the Federal Emergency Management Agency maps.
2. Zone B: Areas located one-half mile upgradient from Zone A or to the watershed boundary, whichever is less.
3. Zone C: Areas encompassing the remaining watershed not designated Zone A or B.

Additionally, two regions overlay the watershed:

- Region 1: The reservoir pool area and Zone A adjacent to the reservoir, as well as areas of potentially high risk as a result of intense development, poor soils, steep terrain, bordering wetlands, and tributary systems.
- Region 2: Areas encompassing the remaining watershed not designated as Region 1.

The basis of the criteria for the Wachusett plan was the travel time of the pollutant to the reservoir and, consequently, the time available for the pollutant to be trapped, adsorbed, diluted, or degraded.

In general, immediate threats dictate remedial action or more aggressive control measures; potential threats will require protective and preventive types of controls. Each threat was rated as high, moderate, or low with general consideration of the following factors related to potential pollutants.

- **Attenuation/assimilation:** The ability of the pollutants to be attenuated through assimilation in the land or water regimes (dilution) is considered. For example, sediment from erosion may be removed from raw water before water is withdrawn or treated. Sediment may be removed by using long detention times, allowing sedimentation within a reservoir; oil and grease from parking lot runoff tends to accumulate without attenuation. However, some water-soluble constituents do not attenuate by settling.
- **Human health effects:** The pollutant's relative threat to human health is considered. For example, nutrients from fertilizer use may promote eutrophication (e.g., algal growth, oxygen depletion) that may in turn affect the aesthetic quality of the water; algae in eutrophic water may even release toxins into the water source; carcinogenic organic solvents from a large spill may pose severe health risks.
- **Frequency of occurrence:** The relative frequency of releases of a pollutant is considered. For example, failing septic systems produce virtually continuous releases of pollutants; urban runoff is frequent but associated with specific periodic events. In contrast, accidental spills on a roadway are very infrequent events.
- **Potential quantity of pollutant:** The relative quantity of pollutant typically released by an occurrence of the threat is considered. For example, quantities of oil and grease released in parking lots at recreational facilities would generally be small, whereas quantities of gasoline leaking from underground storage tanks could be considerably larger.

Table 7-3 Threat assessment

Threat	Zone A		Zone B	
	Region 1	Region 2	Region 1	Region 2
Human Induced				
Onsite septic systems (NPS)	**H**	**H**	**H**	M
Recreational activities (NPS)	M	M	L	L
Stormwater runoff (NPS)	**H**	L	**H**	L
Logging on Missouri Department of Conservation land (NPS)	L	L	L	L
Logging on private land (NPS)	M	M	L	L
Herbicides, pesticides and fertilizers (NPS)	M	M	M	M
Animal grazing (NPS)	M	L	M	L
Construction activities (NPS)	H	M	M	L
Gravel mining (NPS)	L	M	M	L
Road salting (NPS)	M	L	M	L
Future developments (NPS)	H	H	H	H
Uncontrolled releases from transportation activities (PS)	H	H	H	L
Uncontrolled releases from fixed sites (PS)	H	L	M	L
Barnyards and feedlots (PS)	M	L	M	L
Gasoline/petroleum storage (PS)	**H**	H	**H**	M
Permitted activities (PS)	H	L	H	L
Unauthorized activities (PS/NPS)	L	L	M	L
Natural				
Erosion (NPS)	M	M	L	L
Wildlife (NPS)	**H**	M	L	L

Legend:

H = high existing	H = high potential
M = moderate existing	M = moderate potential
L = low existing	L = low potential
NPS = nonpoint source	PS = point source

An example of how the method was applied for the Wachusett Reservoir watershed in Massachusetts is shown in Table 7-3.

In addition, the Massachusetts's Watershed Protection Act defines two additional protection zones:

1. Primary Protection Zone—the area 400 ft (122 m) from the edge of the reservoir and 200 ft (61 m) from tributaries and surface waters
2. Secondary Protection Zone—the area between 200 and 400 ft (61 and 122 m) from the banks of tributaries and surface waters, on land within flood plains, over some aquifers, and within bordering vegetated wetlands

Any generation, storage, disposal, or discharge of pollutants are prohibited in the Primary Protection Zone. In the Secondary Protection Zone, certain activities are specifically prohibited including outdoor, uncovered storage of manure. In certain instances, fecal source tracking has been applied within the Wachusett Reservoir watershed to provide landowners with scientific evidence that certain activities contribute to microbial

Table 7-4 Wachusett Reservoir watershed 2013 public access policy summary

Activity	Intake Protection Zone				
	East Side (Non-reservoir) Route 70 From Gate 5 to Promenade	West Side (Reservoir) Route 70 From Gate 5 to Promenade	Promenade	Road Below Dam and Spillway to Top of Dam	North Dike From Spillway to Area Just Before Gate 36
Off-Road Driving (ORVs, ATVs)	⊘	⊘	⊘	⊘	⊘
Snowmobiling	⊘	⊘	⊘	⊘	⊘
Bicycling	⊘	⊘	⊘	✓	✓[1]
Walking/Hiking	✓[7]	⊘	✓	✓	✓
Dog Walking	⊘	⊘	⊘	✓[2]	⊘
Cross-Country Skiing	✓[7]	⊘	⊘	✓	⊘
Shoreline Fishing	⊘	⊘	⊘	✓[3]	⊘
Fishing with Waders	⊘	⊘	⊘	⊘	⊘
Horseback Riding	⊘	⊘	⊘	⊘	⊘
Hunting	⊘	⊘	⊘	⊘	⊘
Boating – Non-motorized	⊘	⊘	⊘	⊘	⊘
Boating – Motorized (including "jet skis")	⊘	⊘	⊘	⊘	⊘
Swimming	⊘	⊘	⊘	⊘	⊘
Ice Skating/Ice Fishing	⊘	⊘	⊘	⊘	⊘
Camping	⊘	⊘	⊘	⊘	⊘
Picnicking	✓[4.7]	⊘	⊘	✓[4]	✓[4]
Fires & Cooking	⊘	⊘	⊘	⊘	⊘
Programs/Assemblies	✓[5.7]	⊘	✓[5]	✓[5]	✓[5]
Trail Clearing/Trail Marking/Advertising	⊘	⊘	⊘	⊘	⊘
Collecting/Metal Detecting	⊘	⊘	⊘	⊘	⊘
Geocaching/Letterboxing	✓[6.7]	⊘	⊘	✓[6]	✓[6]

✓ - Public access is allowed in designated areas only ⊘ - Activity prohibited

1. Allowed on rail bed from gate 39 to spillway only. 2. Dogs only allowed to top of spillway and must stop at sign. Dog waste must be picked up and disposed of properly. 3. Fishing allowed from fountain impoundment to approximately 275 feet after concrete abutments of old railroad bridge. 4. Trash must be carried off of DCR property. 5. DCR special permit required. 6. Geocaching guidelines must be adhered to and all DCR rules apply. 7. Except for property between Rte 62 (Willow Road) and Mile Hill Roads.

Source: *Massachusetts DCR 2013, 2013 Watershed Protection Plan Update, Chapter 6: Wachusett Reservoir Control Programs*

contamination. Such data can sometimes eliminate speculation and erroneous contamination sources.

Public access and recreational use is allowed in portions of the Wachusett watershed even though this increases the potential for water quality problems. A program of public access management is necessary to permit public access while still providing adequate water supply safety and security. The *2013 Watershed Protection Plan Update* (Massachusetts DCR 2013) provides guidance for public activities on Massachusetts DCR properties. See Table 7-4 and Figure 7-9 for a summary of the plan's policies.

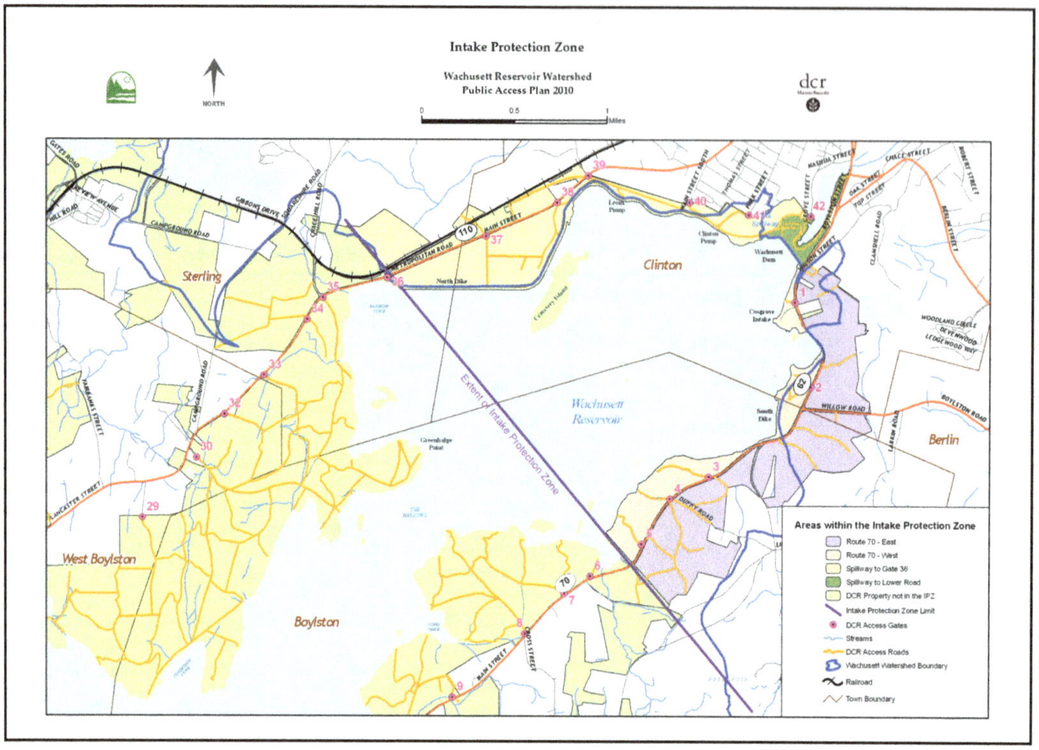

Source: *Massachusetts DCR 2013, 2013 Watershed Protection Plan Update, Chapter 6: Wachusett Reservoir Control Programs*

Figure 7-9 Wachusett Reservoir watershed public access management zones

The plan is reviewed regularly and permitted activities are evaluated periodically to ensure that water quality is not threatened.

Additional guidance on evaluating threats may be found in the *Operational Guide to AWWA Standard G300, Source Water Protection* (Sham et al. 2010).

Rank Threats

It is important to rank threats to the source water quality so those threats with the highest ratings can be controlled first. The WPP should include a schedule for implementing control measures based on the threats with the highest rating and available resources. Immediate threats should always be ranked higher than potential threats (i.e., Zone A higher than Zone B or C, and Region 1 higher than Region 2).

Identify Control Measures to Reduce or Eliminate Threats

The WPP should include measures designated to control specific threats, as well as broader control measures that provide comprehensive protection measures for a number of threats. Control measures are activities used in the watershed to remediate immediate threats and prevent potential ones. Activities with high threats (i.e., high risk), as identified in the ranking step, will require controls that can be implemented easily and reliably. Structural and nonstructural control measures should be considered in the WPP. Nonstructural controls are behavioral modifications that may be developed on the basis of legislation, regulations, and policies. Nonstructural controls can be highly variable and they should be analyzed to evaluate their adequacy in protecting the source water:

- Do these nonstructural measures provide adequate controls for the specific threats identified earlier?
- What new nonstructural control measures should be considered for each unique watershed?

Important considerations about nonstructural control measures include the following:

1. Who is responsible for administering and enforcing the control measures?
2. Does the administrating agency have enforcement powers?
3. Is the proposed control measure effective for the target pollutants?

The structural control measures are physical controls that once put into place do not depend on behavioral compliance to be effective. Examples of structural control measures might include

- detention or retention basins,
- filtration basins and trenches,
- grass swales,
- filter buffer strips,
- porous pavement,
- oil/grit separators,
- containment systems,
- onsite treatment/disposal systems,
- sediment and erosion control measures, or
- animal controls, such as barriers and fencing.

These controls are physical in nature and are targeted for defined areas in the watershed and specific threats or types of contaminants. Most controls are associated with stormwater runoff and function to preclude direct entry of the runoff into the source waters or to mitigate the effect of the runoff.

Quantify the Efficiency and Effectiveness of Control Measures

A matrix approach can be used to match available control measures to specific threats or contaminants. The cells of the matrix may be classified as low, medium, or high with respect to the efficiency of controlling the specific contaminant. *Efficiency* measures the relative removal capacity of the control measure. For example, a detention basin may be highly efficient in removing incoming suspended sediment but only moderately efficient in removing colloidal organic compounds. *Effectiveness* measures the ability of the control measure to remove or mitigate the target contaminants. In this example, the detention pond would be inefficient and ineffective if the target contaminant was dissolved organic compounds. Furthermore, it is important to evaluate these control measures for various types of events. For example, a structural device may be efficient for low-frequency events (e.g., high flows) in comparison to its performance during high-frequency events (e.g., low flows).

Identify Administrators of Control Measures

The effectiveness of various control measures depends largely on the responsible administrative agency. Many control measures are best administered at the local level; however, many local agencies are stressed for resources and cannot always provide the needed maintenance and enforcement resources. In addition, a watershed may cross several jurisdictions, and resources must be coordinated. Therefore, adequate funding for proper administration or coordination is necessary for a successful WPP. Funding may be available from the USEPA Drinking Water State Revolving Fund, USEPA Clean Water State Revolving Fund, USEPA Clean Water Act Section 319 Program, US Department of Agriculture (USDA) Environmental Quality Incentives Program, USDA River Network grants, and USDA Emergency Conservation Program. Water systems should meet with regional USEPA, Natural Resources Conservation Service, and state agencies to determine the availability of funding in their regions.

Rank Control Measures

To rank the various control measures, utilities must first identify and rank the target pollutants or threats. Next, they must determine the removal or mitigation capability of the control measures for the various target pollutants along with the task and cost of implementation. These criteria result in a general protection philosophy and ranked set of protection measures. The protection philosophy exemplifies the distinctive protection approach that can most effectively achieve the objectives of the WPP. For example, the philosophy adopted for the Wachusett Watershed Protection Plan (Massachusetts DCR 2013) resulted in the following ranked actions:

1. Land acquisition is the best and most desirable method for watershed protection.
2. Current regulatory frameworks are used to the maximum extent possible.
3. Enforcement of regulations is emphasized.
4. Environmental management activities that traditionally have been local responsibilities (i.e., local bylaws and enforcement) are supported as the highest priority.
5. Local cooperation and support of public education, financial assistance, and technical support is promoted.

This approach resulted in identifying and ranking control measures for specific threats (shown in Table 7-2).

Develop a Watershed/Reservoir Protection Plan

Although a partnership must be formed to effectively develop and implement the WPP, a lead agency (or lead person of the WPP) can usually be identified. This agency will have primary responsibility for watershed protection, usually including land management and protection, water quality monitoring, enforcement of rules and regulations, recreation, sanitary surveys, and policy making and planning. Other agencies and parties play important roles in developing a WPP. State agencies, for example, are responsible for various environmental regulatory and permitting programs to control land-use activities, and local agencies have an array of local land-use regulations that affect water quality. Also, several federal regulations promulgated by the USEPA may apply to the WPP. The WPP involves

1. developing a vision or the desired future condition for the watershed;
2. delineating the geographic and political boundaries of the watershed;
3. characterizing the threats to degradation of the source water quality and their ranking;
4. developing an action plan that
    - identifies water quality and programmatic goals and objectives,

- identifies effective measures to control or mitigate threats and their ranking, and
- identifies available funding and resources for support of the program;

5. developing a monitoring and assessment program to measure the results of implementing the plan and continually provide feedback for modifying and improving the program.

Implement the Plan

The WPP is implemented based on the ranking of the protection or control measures. Generally, protection measures developed for high-priority threats are implemented immediately because they are the most critical for protecting the watershed. Protection measures for moderate-priority threats are next implemented, possibly over a period of years. Finally, the low-priority measures are scheduled for implementation when financial resources are available. Additional guidance on components and development of implementation plans may be found in the *Operational Guide to AWWA Standard G300, Source Water Protection* (Sham et al. 2010).

The USEPA has identified nine essential elements for an effective watershed-based plan (USEPA 2008) including the following:

(a) An identification of the causes and sources or groups of similar sources that will need to be controlled to achieve the load reductions estimated in this watershed-based plan.

(b) An estimate of the load reductions expected for the management measures described under paragraph (c) below.

(c) A description of the nonpoint source (NPS) management measures that will need to be implemented to achieve the load reduction estimated under item (b) above.

(d) An estimate of the amount of technical and financial assistance needed, associated costs, and/or the sources and authorities that will be relied upon to implement this plan.

(e) An information/education component that will be used to enhance public understanding of the project and encourage their early and continued participation in selecting, designing, and implementing the NPS management measures that will be implemented.

(f) A schedule for implementing the NPS management measures identified in this plan that is reasonably expeditious.

(g) A description of interim, measurable milestones for determining whether NPS management measures or other control actions are being implemented.

(h) A set of criteria that can be used to determine whether loading reductions are being achieved over time and substantial progress is being made toward attaining water quality standards, and if not, the criteria for determining whether this watershed-based plan needs to be revised or, if a NPS total maximum daily load (TMDL) has been established, whether the NPS TMDL needs to be revised.

(i) A monitoring component to evaluate the effectiveness of the implementation efforts over time, measured against the criteria established under item (h) immediately above.

Watershed-based plans that include these nine essential elements are now commonly referred to as nine-element plans. Such a plan is required for funding of watershed projects under USEPA's Section 319(h) NPS management grant program. As noted among the elements included, nine-element plans are broader than source water protection plans. Nevertheless, the actions outlined in a nine-element plan are in parallel with elements of an SWPP and thus should be integrated in an SWPP whenever appropriate.

The nine elements fit nicely into the six essential components of an SWPP identified in AWWA's G300 standard. Element (a) is a component under the G300 "Source Water Characterization" element. Elements (b) through (g) are part of the "Action Plan" element in G300. And finally, elements (h) and (i) are part of the "Evaluation and Revision" element of G300.

Although the WPP is developed and implemented according to a timeline, the plan should be dynamic and be continually updated and modified based on the ongoing monitoring results. Activities in the watershed will change with growth and land-use development; new control technologies will evolve. The partnership needed to make the WPP succeed is a life-long relationship.

REFERENCES

AWWA (American Water Works Association). 2014. ANSI/AWWA Standard G300. *Source Water Protection*. Denver, Colo.: AWWA.

Eberts, S.M., M.A. Thomas, and M.L. Jagucki. 2013. *Factors Affecting Public-Supply-Well Vulnerability to Contamination: Understanding Observed Water Quality and Anticipating Future Water Quality*. USGS Circular 1385. Reston, Va.: US Geological Survey.

Hill, C. 2004. Regulatory Federalism: A Comparative Study of Safe Drinking Water Policy Implementation in Vancouver, Canada, and Seattle, USA. Paper presented to the Canadian Political Science Association, June 5, 2004.

Keith, S., and M. Howes. n.d. Protecting the Source of Your Drinking Water. Iowa Department of Natural Resources and Iowa Geological and Water Survey. www.igsb.uiowa.edu/Mapping/sourceh20/source.htm (accessed May 12, 2014).

Massachusetts DCR (Department of Conservation and Recreation). 2013. *2013 Watershed Protection Plan Update*. www.mass.gov/eea/docs/dcr/watersupply/watershed/2013dcrwatershedprotectionplan.pdf (accessed May 1, 2014).

Massachusetts DEP (Department of Environmental Protection). 1999. Massachusetts Source Water Assessment Program Strategy. www.mass.gov/eea/docs/dep/water/drinking/alpha/i-thru-z/swapstra.doc (accessed May 12, 2014).

MDH (Minnesota Department of Health). 2009. Wellhead Protection Area Delineation Fact Sheet. Drinking Water Protection Section, Source Water Protection Unit. Saint Paul, Minn.: MDH.

NEIWPCC (New England Interstate Water Pollution Control Commission). 1996. *Source Protection. A Guidance Manual for Small Surface Water Supplies in New England*. Lowell, Mass.: NEIWPCC.

NEIWPCC (New England Interstate Water Pollution Control Commission). 2004. *Protecting Drinking Water Sources in Your Community: Tools for Municipal Officials*. Lowell, Mass.: NEIWPCC.

Sham, C.H, R.W. Gullick, S.C. Long, and P. Kenel. 2010. *Operational Guide to AWWA Standard G300, Source Water Protection*. Denver, Colo.: AWWA.

TAMU (Texas A&M University). 2014. Hydrologic Modeling Inventory. http://hydrologicmodels.tamu.edu/ (accessed April 25, 2014).

USEPA (United States Environmental Protection Agency). 1987. *Guidelines for Delineation of Wellhead Protection Areas*. EPA 44016-87-010. Washington, D.C.: USEPA, Office of Groundwater Protection.

USEPA (United States Environmental Protection Agency). 1988. *Developing a State Wellhead Protection Program: A User's Guide to Assist State Agencies Under the Safe Drinking Water Act*. EPA 440/6-88-003. Washington, D.C.: USEPA, Office of Groundwater Protection.

USEPA (United States Environmental Protection Agency). 1989. *Wellhead Protection Programs: Tools for Local Governments*. EPA 440-6-89-002. Washington, D.C.: USEPA, Office of Water.

USEPA (United States Environmental Protection Agency). 1991. *Managing Groundwater Contamination Sources in Wellhead Protection Areas: A Priority Setting Approach*. EPA 570/9-91-023. Washington, D.C.: USEPA, Office of Groundwater and Drinking Water.

USEPA (United States Environmental Protection Agency). 1992. *Final Comprehensive State Groundwater Protection Program Guidance*. EPA 100-R-93-001. Washington, D.C.: USEPA.

USEPA (United States Environmental Protection Agency). 1993. *Wellhead Protection: A Guide for Small Communities, Seminar Publication*. EPA 625/R-93/002. Washington, D.C.: USEPA, Office of Research and Development and Office of Water.

USEPA (United States Environmental Protection Agency). 1997a. *Delineation of Source Water Protection Areas, A Discussion for Managers; Part 1: A Conjunctive Approach for Groundwater and Surface Water*. EPA 816-R-97-012. Washington, D.C.: USEPA, Office of Water.

USEPA (United States Environmental Protection Agency). 1997b. *Guidelines for Wellhead and Springhead Protection Area Delineation in Carbonate Rocks*. EPA 904-B-97-003. Atlanta: USEPA, Groundwater Protection Branch, Region 4.

USEPA (United States Environmental Protection Agency). 1997c. *State Source Water Assessment and Protection Programs: Final Guidance*. EPA 816-R-97-009. Washington, D.C.: USEPA, Office of Water.

USEPA (United States Environmental Protection Agency). 2006. *How-to Manual: Update and Enhance Your Local Source Water Protection Assessments*. EPA 816-K-06-004. Washington, D.C.: USEPA, Office of Groundwater and Drinking Water.

USEPA (United States Environmental Protection Agency). 2008. *Handbook for Developing Watershed Plans to Restore and Protect Our Waters*. EPA 841-B-08-002. Washington, D.C.: USEPA. water.epa.gov/polwaste/nps/upload/2008_04_18_NPS_watershed_handbook_handbook.pdf (accessed May 1, 2014).

USEPA (United States Environmental Protection Agency). 2009. *National Primary and Secondary Drinking Water Regulations*. EPA 816-F-09-004. Washington, D.C.:USEPA, Office of Groundwater and Drinking Water.

USEPA (United States Environmental Protection Agency). 2014a. Potential Contaminant Source Inventory Tools Links. water.epa.gov/infrastructure/drinkingwater/sourcewater/protection/potentialcontaminantsourceinventorytoolslinks.cfm (accessed May 11, 2014).

USEPA (United States Environmental Protection Agency). 2014b. CCL and Regulatory Determinations. water.epa.gov/scitech/drinkingwater/dws/ccl/index.cfm (accessed April 25, 2014).

USEPA (United States Environmental Protection Agency). 2014c. Unregulated Contaminant Monitoring Program. water.epa.gov/lawsregs/rulesregs/sdwa/ucmr/index.cfm (accessed April 25, 2014).

USEPA (United States Environmental Protection Agency). 2014d. Superfund Information Systems. www.epa.gov/superfund/sites/cursites/ (accessed April 25, 2014).

Water Policy and Governance Group. n.d. Governance for Source Water Protection in Canada. www.governanceforwater.ca/ (accessed April 25, 2014).

WDNR (Wisconsin Department of Natural Resources). 2012. State of Wisconsin Wellhead Protection Program Plan for Public Water Supplies, amended May 2012. dnr.wi.gov/topic/Groundwater/documents/pubs/whpp_plan.pdf (accessed April 25, 2014).

Winter, T.C., J.W. Harvey, O.L. Franke, and W.M. Alley. 1998. *Groundwater and Surface Water—A Single Resource*. USGS Circular 1139. Denver, Colo.: US Geological Survey.

Chapter 8

Water Supply Assessments

The choice of either one or an entire portfolio of water supply alternatives for further study or implementation depends on the outcome of an assessment of water supply alternatives and how information from the assessment is used in the decision-making process described in chapter 9, Alternatives Analysis. A sound assessment of water supply alternatives is initiated with a comprehensive list of options as described in chapter 4 (Water Resources Alternatives) and uses the information described in chapters 5 (Water Demand Forecasting), 6 (Water Quality), and 7 (Source Water Protection) to define the risk and sustainability of existing sources and future alternatives. The intent of this chapter is to provide insight into the information needed and approach for assessing the characteristics of various water supply alternatives.

OVERVIEW OF ASSESSMENT APPROACH

Alternative water supply sources are usually evaluated in phases to minimize time, effort, and rework. To avoid detailed investigations of "fatally flawed" alternatives, early in the assessment process the evaluator can conduct a "screening-level" evaluation to generate a list of viable alternatives to carry forward for more detailed investigations. Ultimately, the goal of the water supply assessment is to generate a matrix summarizing the characteristics for each alternative based on a defined set of evaluation criteria. Using the evaluation criteria and working closely with stakeholders, the evaluator's objective through the assessment process is to focus the investigations on those alternatives that are likely to generate a reliable, diverse, and flexible portfolio of water supply alternatives. In the early stages of the assessment, it may be possible for similar alternatives (e.g., several dams and reservoirs) to be grouped and a single alternative can be carried forward as a surrogate or representative alternative in the evaluation process. Thorough documentation of the outcome from each alternative can enhance credibility in the assessment process and save time and money in future stages of study. The outcome of the assessment of alternatives can vary greatly depending on the study objectives, but several possible outcomes include the following:

- *Baseline* is an alternative selected at the assessment outset as a component of the recommended plan regardless of any other alternatives selected (drought management plan, water use efficiency program, etc.).
- *Fatally flawed* or *highly unlikely* is an alternative that was eliminated from further consideration because of lack of reliability, major permitting obstacles, and so on.
- *Backup* is an alternative that is less desirable when compared to the remaining alternatives but will be considered as a potential component of the plan.
- *Cornerstone* is an alternative in the "short list" that could satisfy the entire need and is receiving further consideration as a key element of the plan.

To clarify, chapter 9 (Alternatives Analysis) focuses on various techniques for comparing alternatives, whereas this chapter identifies the assessments and documents the findings needed to conduct the comparisons.

IDENTIFICATION OF EVALUATION CRITERIA FOR ASSESSMENTS

After establishing there is a need for the project and identifying a comprehensive list of potential alternatives, the next step in the planning process is to identify the criteria to be used to characterize the alternatives. A list of evaluation criteria is presented in this chapter to provide guidance in applying the assessment process. The evaluator may wish to engage a stakeholder group in a collaborative process to identify criteria and generate concise definitions for each. The criteria that follow are fairly standard, and the evaluator will likely need to add or alter the criteria when this method is applied for a specific utility. There are, however, some general rules when developing criteria that should be adhered to according to principles of multi-attribute decision making:

- *Non-redundant* means that there should be no overlap between the criteria (i.e., avoid using two criteria that will ultimately measure the same thing).
- *Easily understood* means that the definition and interpretation of the criteria should be easily understood by multiple audiences and stakeholders.
- *Measurable criteria* are not useful if they cannot be measured quantitatively or through an agreed-upon qualitative scoring.
- *Differentiation criteria* should be selected only if they can show some difference between alternatives (i.e., if all the alternatives meet safe drinking water standards and that is the basis for the evaluation, then they will score the same).
- *Concise in numbers* is important because having too many criteria will result in ranking scores that are not significantly different from each other. Generally, limiting the primary criteria to about six or less is preferable.

It is critically important that the evaluation criteria be well defined and clearly understood by the decision makers.

The following water supply criteria are frequently used in the water supply alternatives assessment process. The evaluator (and possibly stakeholders) should review these criteria and tailor the criteria and definitions to meet the needs of the study and the technical, financial, and environmental aptitude of the stakeholders.

- **Reliability.** This criterion evaluates the ability to meet current and future water supply needs with a high degree of reliability. The criterion will measure
  - ability of supply to meet current and future needs,
  - hydrologic and hydrogeologic variability,
  - vulnerability to effects of emergency disruptions, and
  - ability to extend the adequacy of existing supplies (water conservation, water use efficiency, etc.).
- **Water quality.** This criterion evaluates the ability to provide safe drinking water through developing and using water supplies with the best available source water quality for the proposed use. The criterion will measure
  - existing source water quality,
  - vulnerability of potential source contamination,
  - water quality compatibility with other supplies, and
  - water quality preferences.
- **Environment.** This criterion evaluates the ability to minimize effects to the natural environment from supply development through mitigation and other methods, and to seek opportunities to enhance the natural environment. The criterion will measure
  - net environmental benefits, including energy;
  - environmental enhancement opportunities beyond mitigation, including recreation (e.g., boating, fishing, sightseeing); and
  - ecological effects (benefits), such as the extent to which projects damage (or create) important habitats, or create green space and recreation opportunities.
- **Implementation.** This criterion evaluates the ease of, or ability to, implement a source of supply. This criterion will measure
  - ease of obtaining permits (especially water rights),
  - effects on tribal treaty interests,
  - public and political support,
  - institutional barriers,
  - project timing and readiness,
  - ability to spread costs over time,
  - ability to expand beyond planning period,
  - ability to accommodate the entrance or exit of members from regional agreements, and
  - ability to maintain local ownership.
- **Cost.** This criterion evaluates the overall unit cost (e.g., $/mgd) of an alternative and reviews
  - capital,
  - operation and maintenance,
  - present worth,
  - life-cycle costs,
  - treatment or operational functions that rely on natural systems for lower life-cycle operating costs through reduced energy and chemical inputs,
  - energy source including reduced reliance on the energy grid through greater energy efficiency or self-generation of energy,

- the extent to which projects focus on sustainability of infrastructure in a utility's existing service area, and
- flexibility of option to be idled from year to year with limited financial effects.

The elements of the criteria included in an evaluation—and the level of detail associated with the investigations for each alternative—are dependent on the specific alternative being considered and the nature of the study. It is important that the level of detail associated with the elements be appropriate to produce an "apples to apples" comparison of alternatives. In addition, some of the elements listed above, such as environmental mitigation, may be viewed as "benefits" by some participants and "costs" to others. To measure the effectiveness of the criteria, specific metrics and their location should be established for each element and could include items such as reservoir pool levels, number of days of acceptable flow for whitewater rafting, water quality characteristics, and so forth. Again, these metrics can be developed in collaboration with stakeholders, regulators, and other participants.

WATER SUPPLY RELIABILITY ASSESSMENT

Reliability and Risk

Water supply reliability reflects the ability of an alternative to meet current and future municipal water supply needs. Public drinking water utilities are expected to achieve a high level of reliability delivering sufficient, high-quality drinking water 24 hours a day, 365 days a year. While 100 percent reliability is not always possible because of extenuating circumstances (extreme weather, human error, accidents, and other unforeseen events), it is the goal. Therefore, it is necessary to assess the ability of the source (or sources) of supply to consistently meet the utility's intended need in terms of supply, quality, and infrastructure (e.g., treatment and delivery) at a reasonable cost.

A general approach for assessing reliability includes establishing system reliability without options and strategies (baseline), defining the vulnerability of the existing supply system, and computing reliability with improvements in place. When evaluating new sources of supply, it is inherently a long-term planning decision because it is implicit that the new or expanded source of supply will be used by the community in perpetuity. Long-term decisions involve risk because the future conditions cannot be known with 100 percent certainty. Therefore, it may be beneficial to assess the risk that supply options may not perform adequately under various conditions, may be too expensive, or may face community opposition.

To accommodate incremental growth of populations, coupled with the cost and difficulty of constructing single large-scale water projects, many utilities are pursuing a portfolio approach to water supply planning to minimize risk. The portfolio approach, initially developed for managing financial resources, focuses on targeting multiple potential sources (groups) for development. The portfolios define the order in which diverse options would be implemented when responding to emerging system vulnerabilities. The added benefit is that a mix of resources enables the utility to hedge against future uncertainty. A more detailed discussion of uncertainty is included in chapter 9, Alternatives Analysis.

Hydrologic Reliability

Hydrologic reliability reflects the ability of the resource to supply a sufficient quantity of water to meet the current or projected need when it is required. For simplicity, *hydrologic reliability* is referred to as "yield" and is typically defined for surface water supplies as the *maximum rate of withdrawal that can be sustained based on the worst drought in the observed hydrologic record*. Yield for groundwater supplies may be determined based on withdrawal rates through the drought of record, or it may be determined from pumping tests to estimate the maximum yield that does not result in excessive drawdown of the aquifer.

The first step in assessing the reliability of water supply alternatives is establishing the need for the project (i.e., quantity of deficit or shortfall in supply in a specific year). The need for an alternative source of supply is the foundation for justifying the project to regulators, stakeholders, and the general public. A "Needs Statement" is the cornerstone of many permitting processes. As such, the need must be clearly defined and understood by various audiences. Consequently, the demand condition (average annual, seasonal, emergency, etc.) and the characteristics of the supply source must be carefully considered. Supply sources that include raw water storage reservoirs are typically sized based on average-day demands, whereas withdrawals from river supplies are typically sized for peak-day demands.

The capabilities and reliability of alternatives in terms of water supply can be determined based on the configuration of the alternative and modeling. The following hydrogeologic investigations are most commonly used for groundwater quantification:

- Analyze hydrologic data for water budget (rainfall, streamflow, subsurface flows, boundary conditions, spring flows, land use, water use, irrigation-applied water, evapotranspiration).
- Correlate and integrate various sources of data, such as water-level maps, well hydrographs, geologic maps, well logs, geophysical logs, driller's logs, geochemical analyses, river flow analyses at various gauging stations in the groundwater basin, pumping tests, observations at monitoring wells, isotope methods, and observations from test holes.
- Conceptually model the groundwater basin using a standard modeling approach.

For surface water sources, hydrologic models are typically used to evaluate various layouts and operating scenarios for potential water supplies. Depending on the utility's needs, the model can be a simple spreadsheet driven by readily available streamflow data, or it can be a complex multimodel ensemble that includes meteorological parameters, watershed processes, and groundwater flows. A comprehensive discussion on developing models is beyond the scope of this chapter.

General guidelines for conducting a model development project are listed below. Model development can be an expensive effort, so proper planning is critical at the outset to identify model objectives and understand the level of detail needed (or possible) to establish the level of effort and budget required to achieve the organization's goals.

1. Establish scope (breadth and depth) of the inquiry and investigation.
2. Understand the purpose of the project and how modeling fits within the overall project.
3. Discuss relevant issues and goals of the modeling task with the client.
4. Develop a data management plan and discuss hardware and software compatibility with the client.
5. If possible, involve the client in the model selection process, or at least educate the client on the selection criteria and process.
6. Consider conducting interviews with software vendors (if applicable).
7. Discuss assumptions and limitations of the model with the client.
8. Consider the costs of hardware and software, as well as the cost for needed or desired refinements of a model.

9. Start compiling and collecting data as early as possible in the model development process.
10. Develop a communications plan (i.e., determine what raw model results will be post-processed and presented in graphical or tabular forms, in a format that is understandable by decision makers, and what level of reporting will be published; if possible, prepare a report outline with key details as early in the project as possible).
11. Define project milestones, for example
    - complete data collection,
    - complete input data preparation,
    - develop a preliminary hydrologic budget,
    - initially test the model with input data and review the first round of results,
    - finalize the hydrologic budget and first round of calibration,
    - identify the second and third rounds of calibration and reassess calibration goals,
    - finalize the model calibration and customer buy-in of the model,
    - formulate alternatives to be studied using the model,
    - develop a modeling approach for alternatives and prepare input data,
    - present and have client review results of alternatives analyses,
    - submit the draft report, and
    - respond to client comments, submit a final report, and close out the project.

It is critical that a modeler or evaluator understand the potential sources of uncertainty associated with a given alternative. For example, the possible sources of uncertainty that can affect the assessment of the adequacy of supply (i.e., level of risk) for a reservoir-river system are many-fold and may include the following variables:

- *Reservoir operations and travel times.* This pertains to the precision of the valving or releases from a reservoir to meet downstream flow targets or constraints. In addition, travel times to downstream targets can vary depending on the volume of flow in the river.
- *Water withdrawals and demands.* Water demands for utilities vary throughout the day and seasonally, which can make it difficult to predict the quantity of water to release downstream. Demand uncertainty can be factored into the reliability assessment by addressing current projected growth, slow growth, and rapid growth scenarios (see chapter 5, Water Demand Forecasting).
- *Losses underground.* River systems sometimes exhibit significant "loss" of flow underground, and the magnitude of this loss can vary depending on river flow conditions.
- *Inflows from tributary streams.* Localized thunderstorms in the tributary streams to the main-stem river can create the impression of "excess" releases from a reservoir because they can produce flows above the target levels.
- *Changes in return flows.* The difference between the amount of water withdrawn and water returned to the source (or discharge) by a wastewater treatment plant (or other entity) is usually taken to represent "consumptive use." The percentage of return flow from each of the wastewater plant, power plant, and so on, must be defined under future conditions.

- *Accuracy and availability of stream gauge data.* Streamflow gauges must be routinely calibrated and provide accurate and useful information at the desired location.
- *Variability of drought events.* Droughts more severe than the critical drought that occurred in the previous years of record will occur in the future. However, the date of occurrence, magnitude, and duration of these future events are not known.
- *Climate change.* Shifts within the hydrologic cycle due to climate change are expected in the future, but the site-specific effects may not be known at this time.
- *Changes in irrigation withdrawals.* Irrigation withdrawals tend to be highest when conditions are dry. Historic irrigation withdrawals may be accounted for by using actual stream gauge data, and an assumption may need to be made regarding whether the percentage of available water used for irrigation withdrawal will remain unchanged in the future.

The possible sources of uncertainty that affect the assessment of a supply source's adequacy are many-fold and include not only demographics and water use but uncertainty regarding weather, hydrology, accuracy of stream gauging, and many other factors. While the uncertainty of some of these factors can be mitigated, many cannot and therefore must be addressed in some other fashion such as optimizing existing water supplies or developing new supplies.

In summary, the reliability and adequacy of existing supplies, and the need for the project itself, should be established before starting to identify and evaluate potential alternatives. An excellent starting point for identifying yield for existing and new supply sources is a review of previous water resources models and studies.

Infrastructure Reliability

Infrastructure reliability is the ability of the utility to maintain reliable operations of the water supply, treatment, and piping infrastructure to treat and deliver the water to its customers. This includes the transmission pipelines, pumping stations, and treatment infrastructure, in addition to necessary power supplies needed to make the source of supply available for distribution. Inevitably, infrastructure needs maintenance and replacement, and redundancy enables the source to provide supply during planned or unplanned maintenance or upgrades. One of the primary considerations for all systems is the reliability of the power supply. Other considerations include access to facilities and vulnerability to disruptions during extreme weather conditions (i.e., flooding, drought), age of the existing assets, redundancy of key supply facilities, proximity of supply to demand center, and so forth.

When evaluating the level of reliability required of a water supply source, it is necessary to understand the likelihood and consequence of failure from the loss of the supply. *Infrastructure resilience* reflects the ability to reduce the magnitude or duration of disruptive events. The effectiveness of resilient infrastructure depends on its ability to anticipate, absorb, adapt, and recover from an event. A utility with a suitable emergency source of supply that can be used to prevent a shortfall (such as exceeding permitted groundwater withdrawal rates or purchasing water from other utilities or water rights holders) has different consequences than a utility that does not have any alternatives. In the former situation, there may be undesirable consequences of additional costs, penalties, or political fallout; however, in the latter, the consequence may be running out of water. Another factor is the duration of the loss of supply. In some cases, a short-term loss of supply can be weathered using accumulated storage in the distribution system, while in other cases, any loss of supply has the same consequences regardless of duration.

Key Aspects of Reliability Assessment

While an in-depth discussion of the assessment methodology for each type of alternative (e.g., surface water, groundwater) is beyond the scope of this manual, the evaluator may want to consider the following key aspects during the assessment of reliability for the various sources of supply:

- *Rivers.* During extreme droughts, the yield available from river flows may be less than established minimum instream flow targets. Streamflow gauge data extending over a long period is frequently needed to model competing uses (i.e., waste load assimilation, irrigation withdrawals, recreation, etc.) for the river. If the river system is supplied by snowmelt, climate change could affect the yield. Hydrologic modeling will likely be necessary to establish capabilities and uncertainties of existing supply and address competing uses.
- *Reservoirs.* Droughts are predicted to become more severe in the future because of the effects of climate change. Systems that can store more water may be more resilient to droughts. Reservoirs can decline over time because of source degradation, reservoir siltation, changes in permit requirements, and other factors. Hydrologic modeling is frequently used to generate a water budget in the reliability assessments for water supplies, which include reservoirs, particularly those systems having multiple reservoirs or reservoir-river systems. Sophisticated hydrologic models can be constructed to evaluate the sensitivity of the yield to numerous factors, including river flow diversions, climate change, drought triggers, sedimentation, environmental releases, and so forth. Infrastructure reliability can be enhanced when reservoirs are located in upland areas and flow can be provided by gravity.
- *Stormwater.* Capturing stormwater is very similar to river withdrawals with the exception that there is typically no base flow for continuous capture. In most cases, diversion and storage are necessary to maximize the capture (yield) from seasonal storms that can vary widely in duration and intensity.
- *Groundwater.* Historic hydrogeologic studies should be reviewed to define the performance history of existing wells in the vicinity of the proposed new well. Any regulatory restrictions on withdrawals (i.e., groundwater protection areas) should be identified. Consider conducting pumping tests to collect performance data on aquifer and well.
- *Aquifer storage and recovery (ASR).* ASR wells can improve water basin management by storing water underground from periods of excess supply (flood season or wet hydrologic periods), and later allowing a portion of the stored water to be extracted during periods of demand or short supply (such as irrigation season or dry hydrologic periods). There is no evaporation in the aquifer, as would be the case in a surface reservoir, and in some ASR wells, the recovery is very high.
- *Infiltration galleries.* The yield from infiltration galleries beneath a water body is normally much higher compared to the yield from galleries adjacent to the water body. However, constructing infiltration galleries is usually more difficult under a water body.
- *Reclaimed water.* A critical issue associated with using reclaimed water is balancing supply and demand over time. Seasonal flow variations in municipal wastewater generated in a given service area can be substantial depending on the fluctuations in population (e.g., resort communities), weather conditions (wet areas like Florida), and size of the population (smaller communities typically experience greater fluctuations). If imbalances will occur, storage may need to be constructed

to make these supplies reliable (see chapter 4, Water Resources Alternatives). Note that while reclaimed water storage can contribute to supply reliability, it can also provide an occasion for quality degradation. The ability to reliably produce water that meets strict quality standards must be considered.

- *Drought management plan.* The purpose of a drought management plan is to manage a period of declining supply and increasing demand so that demand does not exceed either supply or system capacity. This is a key difference between drought planning and other water resources planning functions that have as their goal the provision of sufficient water supply and system capacity to meet all demands. Drought management is a means for dealing with a temporary supply/demand imbalance. The magnitude and frequency of the shortages that should be dealt with by drought management is a policy decision. The effectiveness of a drought management plan can be evaluated based on modeling and can be dramatic depending on the triggers, specified actions, and the level of participation.

In summary, assessing the reliability of water supply alternatives should provide the evaluator with a sense as to the capabilities and characteristics of the existing water supplies, as well as each water supply alternative. If the assessment is conducted in phases, a preliminary assessment in the initial stage will allow the evaluator to identify those alternatives that are so limited in supply capability that they are considered "fatally flawed" or "highly unlikely" and are not included in the more detailed investigations in latter phases of study.

WATER QUALITY ASSESSMENT

Water quality is as important as water quantity when evaluating the feasibility and long-term potential of alternative water supply sources. For the water alternatives analysis, the criteria for water quality reflects the ability of the water supply source to consistently provide safe drinking water through developing and using water supplies with the best available source water quality for the proposed use.

Water Quality and Treatment of Supplies

When assessing the merits of water supply alternatives based on water quality, it is important that the water be of sufficient quality for use by the utility based on the treatment technologies employed. Groundwater supplies tend to be relatively stable in terms of water quality compared with surface water supplies that may vary based on seasonal changes or extreme weather. It is therefore necessary to understand the potential variations expected over the life of the water supply source, not only to design treatment technologies but also to understand the frequency that water quality excursions might occur. While it is unclear how average precipitation will change with temperature because of climate change, it is generally agreed that storm severity will probably increase. More intense, severe storms may lead to increased erosion, which will increase turbidity in surface waters. Areas that already observe water quality responses to rainstorm intensity may be especially vulnerable.

Source water quality reliability and treatment plant performance should be evaluated together. The production capabilities of a treatment facility and the withdrawals from the source serving the treatment facility may need to be curtailed if the quality of the source degrades under certain hydrologic events, such as drought. In other cases, for example with offline reservoirs, diversions into the reservoir may be temporarily curtailed if source water quality is poor. The cost of treatment and availability of supply must be evaluated to understand the trade-off between more costly, but robust, treatment with more regular supply reliability. For utilities with multiple sources, the utility must decide whether it is necessary to apply more robust treatment technologies or it is acceptable to blend supplies or take a supply offline temporarily to save on capital or operation and maintenance costs.

Water treatment processes employed are, in most cases, able to operate under a wide range of raw water quality conditions. However, production rates may decrease and treatment costs may increase as water quality worsens because of decreased filter run times, increased chemical use, and increased manpower. Another consideration from a quality perspective is potential future regulatory changes. In evaluating a source of supply, if there are objectionable constituents present but they are below regulatory limits (or can be controlled inexpensively through blending), there is potential that regulatory limits may drop in the future. Consequently, future treatment costs or management strategies must be taken into consideration when considering sources and treatment technology.

Under the constraints of rising energy costs and increasing public and regulatory attention to the environmental impacts of factors like carbon dioxide emissions, the need for comprehensive and defensible approaches to evaluate water supplies and water treatment options is necessary. Carbon footprint models and life-cycle assessments (LCAs) are a means to understand and identify effects associated with various treatment or process options, providing key information that facility managers can use to determine how best to treat, manage, and convey water supplies (see chapter 9, Alternatives Analysis). The LCA analysis can be used in the evaluation of each alternative to

- develop costs for supply or treatment alternatives with similar function (e.g., selection of a primary disinfectant or clarification process);
- quantify the effects of an alternative beyond economic cost; and
- define potential effects due to changing regulations and increased attention to environmental stewardship.

Characteristics of Water Quality

As described in chapter 6, Water Quality, rivers and streams are rich and diverse ecosystems. The topography of rivers and streams affects the velocity, mixing characteristics, solids and sediment transport, and water flow. Lakes and reservoirs are sources for recreation and municipal and industrial uses including agriculture, water quality control, and fisheries management. Lakes and reservoirs can become sinks for nutrients, toxicants, and other substances from incoming rivers; as a result, eutrophication is a significant water quality problem. The drainage area and length, depth, surface area, and volume of a lake or reservoir are significant factors affecting the extent to which the source will act as a sink. Water quality modeling can be conducted using data samples collected as part of a monitoring program.

All the natural and human factors influencing water quality apply to groundwater either directly or through surface water and groundwater interconnections. Groundwater is principally affected by the seepage or leaching of minerals and/or contaminants through soils and by pumping. The evaluator should review data collected to assess the levels of iron, manganese, hardness, radon, volatile organic chemicals, pesticides, and inorganics such as nitrate. High chloride levels have been found near saltwater where groundwater is pumped faster than the water is recharged. If the water table lowers, salt water can intrude.

The water quality merits of the alternatives should be evaluated based on the multitude of regulations established by the USEPA that protect and regulate the quality of water to be used as a source of water supply. Examples include the Clean Water Act,

Stage 2 Disinfectants and Disinfection Byproducts Rule, Groundwater Rule, and Long Term 2 Enhanced Surface Water Treatment Rule. When comparing alternatives, the evaluator should review the following characteristics:

- Physical (total dissolved solids, turbidity, color, tastes and odors, and temperature)
- Chemical (pH, cations, conductivity, and carbon dioxide)
- Biological (bacteria, viruses, protozoa and parasites, and algae and fungi)

A more detailed discussion of these characteristics is presented in chapter 6, Water Quality. Note that the evaluator should include the costs associated with addressing concerns related to these characteristics, (e.g., installation of a reservoir aeration system to address thermal stratification and algae blooms).

Potential Contaminants

When comparing alternatives, the evaluator should consider the following natural and human (point and nonpoint sources) factors that influence the quality of a water source:

- *Natural factors* include climate; watershed characteristics (such as topography, vegetative cover, and wildlife that can significantly affect water quality); geology of a watershed, which can influence erosion, release of chemicals for geologic formations, and so on; potential for fires in the watershed; potential for saltwater intrusion; and reservoir thermal stratification and eutrophication.
- *Point sources* include wastewater treatment plants, industrial facilities, confined animal feeding operations, railways, highways, and gas or oil pipelines. These sources of contamination should be documented to understand the risk of losing the supply source and the potential duration the source may be offline.
- *Nonpoint sources* are runoff or recharge from current and future land use in the watershed, particularly agricultural activities. Recreational activities may affect the biological and chemical components of a water source. Recreational use of proposed water supply reservoirs has become a significant issue with new surface water sources. The biggest issues include swimming and boating (including sizes and types of motors).

Susceptibility assessments are the least understood (and most important) aspects of source water protection. Assessments are intended to identify the contaminant sources that pose the greatest threat to the drinking water supply so they can be targeted for proper management practices. Assessment methodologies include simple, analytical techniques such as hydrogeologic and hydrologic mapping to identify the relative vulnerability of groundwater and surface water supplies, as well as complex contaminant transport models linked to risk assessment matrices. Surface water assessments frequently rely on predictive models.

Sampling and Monitoring of Water Quality

As part of the evaluation of alternatives, the evaluator will typically review water quality data for potential sources or collect water samples for laboratory testing. The frequency, duration, and intensity of water quality measurements depends on the designated use of the source as mandated by federal and state regulations. Continuous sampling and analytical devices are increasingly being used to monitor raw water quality on a continuous basis, also allowing for detection of certain parameters as indicators of accidental spills or releases. These can be useful tools for water utilities. Appropriate agencies should be contacted for regulations once the use of a water source is defined. Statistics are generally used to quantify the quality of sampling results. The numbers and types of samples collected and tested simply to meet regulatory requirements are generally considered a minimal effort in determining water quality.

A sampling and monitoring program must consider the changing activities and conditions in the watershed. Agricultural, recreational, municipal, industrial, and other activities affect downstream water quality. The watershed should be assessed before a final comprehensive sampling and monitoring program is designed and implemented. Potential point and nonpoint sources of contamination should be inventoried. If a monitoring program has been established before such an assessment, the results of the data may be more difficult to interpret.

The weather may be the most unpredictable influence on water quality. Drought conditions result in a reduced dilution factor and may attract wildlife and livestock to the major streams used by public water supplies. The influence of discharge from a waste treatment facility is magnified by a reduced dilution factor during droughts. The effect of climate change and rising sea level may also have an effect on the quantity and quality of freshwater around the world.

More detailed discussions of water quality and source water protection are provided in chapters 6 and 7, respectively.

Key Aspects of Water Quality Assessment

The evaluator may want to consider the following key aspects during the water quality assessment of various sources of supply:

- *Rivers.* Water quality for rivers and streams tends to be more dynamic compared to reservoir sources. Watershed assessments and other studies may be warranted to identify potential sources of contamination (highway crossings, municipal or industrial discharges, pipelines, etc.). In large rivers, it may be desirable to evaluate the best location for a water intake, which may be near the middle of the river to avoid localized influences from tributaries immediately upstream of the intake. Water quality modeling may be warranted to assess the feasibility of the source or to define treatment techniques.
- *Reservoirs.* Lakes and reservoirs can become sinks for pollutants, so eutrophication can be a significant problem. It may be necessary to include an aeration system, or other mixing measures, in the reservoir to improve water quality. It may be prudent to build in the cost for best management practices to address pollutant concerns in the contributing watershed.
- *Groundwater.* Locations of potential contamination sources of aquifers should be identified using source water assessment guidelines. The characteristics of the aquifer under consideration can be defined based on historic water quality data or collecting samples. Infiltration galleries below a water body tend to have higher turbidity and higher bacteria count because of less extensive filtration. Where such withdrawals are considered to directly affect the quantity or quality of surface flows, many of the considerations presented for direct surface water withdrawals will need to be followed, including meeting Surface Water Treatment Rule requirements. This water would likely require filtration, which increases the cost of infiltration galleries. In some locations, assessments of potential for saltwater intrusion should be conducted.

ENVIRONMENTAL ASSESSMENT

The fate of numerous major water supply programs throughout the country has been decided by the outcome of complex and costly environmental assessments. Therefore, it is critical that the evaluator thoroughly understand the requirements and investigations needed to successfully navigate the environmental permitting process. Assessing environmental impacts and benefits addresses how effects to the natural environment can be minimized—from supply development through mitigation and other methods. The assessment can also help seek out opportunities to enhance the natural environment. Environmental documentation should be developed for all projects and discretionary actions that may directly or indirectly change environmental resources or land-use patterns. These could include constructing or modifying structures; operating facilities that affect biological, human, or physical resources; or adopting plans that will affect future land uses. Projects that do not require environmental documentation are generally related to continued administrative or maintenance activities. Emergency projects to protect or restore property generally do not require environmental planning activities.

Environmental Planning Process

The environmental planning process must be implemented concurrently with the technical evaluation, which allows the decision-making processes to be integrated. If the technical process precedes the environmental process and identifies final alternatives that may have significant effects, the cost to redefine and reevaluate new alternatives may be high. In addition, the community may be able to identify the most acceptable alternative. The environmental planning process includes significant public involvement to identify that sensitive community, special interest stakeholders, and resource agencies. Public meetings could be held as part of preparing the technical document. Many stakeholders, however, do not attend the technical meetings because they think they may not understand the information. Therefore, it is important to implement the environmental planning process with the concept development portion of the technical process and present the information to the public in an understandable and timely manner.

The environmental planning process is most successful when the following steps are implemented:

1. *Define the scope.* A critical aspect of scoping is public participation—a step often left until the end of the process when, in reality, it should be part of the scoping mechanism at the beginning of the process. Public participation identifies potentially affected populations and allows for a channel of communication between the project team and the public. Rather than just a public meeting, public participation is an analytical process that can involve interviews, workshops, focus groups, facilitated working sessions, and so on. If federally recognized Native American tribes are part of the public, they should be consulted in a formal government-to-government manner. (See chapter 2 for guidance on effective public participation and consultation.)
2. *Assess the environmental compliance needs.* Identify the agencies that will provide approvals and permits for the final proposed action. Through direct contacts with these agencies, collect their permit applications and submittal requirements, including mandatory requirements for environmental impact documentation. Discuss their approval processes and schedules for approvals. If possible, ask to view examples of successful submittals.
3. *Complete a preliminary analysis of potential issues or constraints.* Review information collected from the agencies with the engineering or economic teams on the

project. Complete preliminary site visits, if applicable, to identify any specific issues related to special habitats, restraints caused by adjacent land uses or utilities, or sensitive noise or visual receptors in the area.

4. *Develop a permitting strategy.* Prepare a list of relevant regulations and requirements, including key schedule milestones. For each milestone, identify required submittals and the information that will be needed to complete the submittals. Coordinate with the engineering or economic teams to develop a critical path for completing the environmental planning process.
5. *Coordinate with all agencies and interest groups.* The most frequent complaints about the environmental planning process are related to delays in receiving approvals or permits and applying mitigation measures or requirements on a project at the end of the project. Both issues usually lead to costly delays and redesign of the proposed alternative. Therefore, frequently coordinating with the agencies and interest groups involved and making them part of the alternative development and evaluation program are important steps. They will likely better understand the reasons for the final decisions.

Types of Environmental Impact Documents

There are three basic types of environmental impact documents: (1) exclusions or exemptions, (2) environmental assessments, and (3) environmental impact statements or reports.

- *Exclusions.* Exclusions are used for projects that occur routinely with anticipated outcomes that do not result in significant adverse effects. Examples include restoring or rehabilitating facilities to a level that does not increase the original design capacity; adding health or safety devices; performing maintenance activities (which may have short-term construction effects); changing ownership of facilities (although the transfer may result in higher costs that may lead to change in lifestyle of users); or converting small structures (although the conversion may result in a change in use). Most of these issues are not considered when preparing water resources plans. Exclusions cannot be used if the proposed action would have precedent for future actions with potential adverse effects. Exclusions also cannot be used if the action would have potential adverse effects on listed species, historic or cultural resources, floodplains, wetlands, water quality, or air quality.

  The lead agency summarizes the reasons for using an exclusion document and adopts the findings. That agency must notify the public of the decision and allow public comment before implementing the proposed action.
- *Environmental assessments.* Environmental assessments or preliminary studies are frequently used in water resources planning. Through this process, a preliminary report is prepared to describe the affected environment or current conditions with respect to the biological, human, and physical resources. The level of detail presented is minimal and focused on the elements that may be affected by the proposed action. Generally, the environmental assessment describes only the potential effects and benefits of the proposed action. If the proposed action could have an adverse effect, it is changed through either the project description or mitigation measures described in the document. Alternatives to the proposed action are described, and the reasons for not pursuing the alternatives are summarized. The environmental assessment is frequently distributed for public comment before the proposed action is adopted, and mitigation measures are included in the final document. If the lead agency determines that there would be no significant adverse effects from implementing the proposed action, including

mitigation measures, the environmental assessment is adopted and recorded in a public forum. The public must be notified that the final environmental assessment has been adopted and that they are allowed to comment on the decision before the proposed action is implemented.

If the environmental assessment shows that significant effects could occur or if the public review requires that several alternatives be evaluated in equal detail, the information in the environmental assessment is used to prepare an environmental impact statement or report. Sometimes an environmental assessment is completed to focus the analysis in an environmental impact document. The important issues are defined by distributing the environmental assessment to the public and agencies. This can limit the level of effort for the actual environmental impact report to the sensitive issues.

- *Environmental impact statements.* Environmental impact statements or reports are used when several alternatives need to be evaluated at an equal level of detail. This type of environmental document includes a detailed description of the affected environment, and potential effects and benefits of alternatives. Frequently, a definition of significant impact is given for each element considered in the report. The adverse effects are measured against effects that would occur without the project and are defined as significant or not significant. The level of analysis of the impact statement is more detailed than under an environmental assessment and frequently includes results of numerical models, field investigations, and discussions with community representatives or experts. The environmental impact documentation also includes an evaluation of cumulative effects of implementing the alternatives with other planned programs that are not specifically part of the alternatives. Mitigation measures are proposed for the alternatives to reduce adverse effects to a level of less significance. The environmental impact document usually defines an environmentally preferred alternative and the preferred alternative. Frequently, these alternatives are identical. However, they may be different, and the document must describe the reasons to select the preferred alternative with fewer benefits to the environment.

  The environmental impact document generally has more stringent public notification requirements during and following report preparation. For example, scoping meetings are required during the preparation of the environmental impact document, as well as public hearings after the draft environmental impact statement is distributed. Additional public involvement allows concerned stakeholders to participate in the process to identify and evaluate alternatives.

For some complex environmental documents, especially for watershed planning, the draft environmental impact document frequently does not include a preferred alternative or final mitigation measures. This process allows the public to help define the preferred alternative and final mitigation measures. In some cases, this approach has allowed stakeholders to define an acceptable approach to sensitive issues.

Critical Steps in Preparing Environmental Documents

In many complex projects, including regional planning, identifying the alternatives is one of the most critical and sensitive steps. With respect to preparing the associated environmental documents, several critical steps—including selecting the study area, study period, and baseline conditions, and determining the level of analytical detail—can determine whether the document will be accepted or legally challenged.

Selecting the study area. In many programs, the study area boundaries are evident because of geographical limitations. However, for watershed programs, the study area boundary (e.g., evaluating how a proposed dam would affect the inundation area and the area along the stream below the dam) could be extremely broad. In a small watershed, the affected area may extend beyond the basin divide because of animal migration. If the stored water or potential hydropower is to be used outside the watershed, the study area may need to be expanded to identify direct economic benefits or effects of operating the dam. If the dam would block anadromous fish migration and this would affect regional fish populations, the study area may need to be expanded to include the geographical boundaries of the fish migration patterns. If the fish form the basis of a commercial or sport fishing industry, the study area may need to be expanded to include the geographical locations of this economic sector.

Expanding the study area will allow stakeholders to participate who do not live adjacent to the proposed project. This will help participants better understand the regional limitations, costs, and benefits, and it may avoid costly litigation or voter challenges. Expanding the study area will increase the cost and time needed to prepare the environmental plan; however, if potential effects are not disclosed in the environmental document, affected parties may successfully litigate.

Selecting the study period and the baseline conditions. Determining the length of the study period and the baseline conditions frequently depends on the type of project being considered in the document. For many construction projects, the study period may be identified as the current conditions. For utilities that provide public services, the study period extends to the projected future when the capacity of the system will be fully used. For watershed projects, especially those with environmental restoration components, the study period may extend 50 years or more into the future. The extent of the study period becomes a critical issue because it defines the baseline conditions on which the effects of implementing the alternatives are compared.

If the study period for a program is shorter than 5 or 10 years, the current conditions or affected environment description can serve as the "No Action" or "No Project Alternative" baseline. For programs that will extend for more than 5 or 10 years, the environmental document must define the baseline conditions for the "No Action" or "No Project Alternative" as the future conditions without the potential alternatives. This becomes difficult when ongoing programs—which must take into consideration changing legislation, weather conditions, and economics—dynamically affect the environment. The environmental analyses need to compare only effects of the alternatives with the future "No Action Alternative" condition. However, many state regulations require that the alternatives to the current and future "No Action Alternative" conditions be compared.

Utilities should also consider the time for critical species populations to react when selecting the study period for environmental restoration projects. For example, if the life cycle of an animal that would benefit from environmental restoration is 5 years, waiting three life cycles (i.e., 15 years) may be necessary to determine whether the restoration activities result in sustainable population improvements.

Determining the level of detail in the analysis. The level of detail for the impact assessment and presentation of the results is also crucial in determining the usefulness of the environmental impact document. Including all available information about a subject in an environmental document is not desirable. The document should refer to information from readily available publications to avoid duplicating other reports. Detailed analyses that need to be completed may be presented in attachments or appendices to the environmental impact document and summarized in the main document. The actual presentation of the data in the main document should be easily understandable by the public. However, information that would usually be reviewed or prepared by a specialist may be presented in the attachments or appendixes. When the information is summarized, the results should not be presented on such a broad basis that the benefits and effects on specific communities or habitats cannot be identified.

As technical reports have become more reliant on computer models, the level of analysis has become more complicated and intricate; therefore, the level of detail must be determined before the analyses are completed. Identifying the level of analysis may be constrained by the available analytical tools or input data. For example, if hydrologic data are limited to a main stream in the watershed, constructing a model to simulate operations of all tributary streams would be more difficult than if data are available for all subbasins.

Before initiating the development of a new analytical tool or use of a tool, the lead agency must determine the level of detail required by considering the following: (1) whether the level of detail is appropriate to discern the differences between alternatives, and (2) if the level of detail is appropriate to disclose potential effects to an individual portion of the study area or community. The level of detail needed to discern the differences between alternatives depends on the facilities or policies considered. For example, if regional watershed management alternatives range from construction of a new dam to stringent conservation of water, providing a surface water analysis that extends to every tributary in the watershed may not be necessary. However, if the alternatives might affect one part of the watershed to a higher degree than other areas because of changes in groundwater use, a groundwater model for the watershed may be needed.

Content of Environmental Impact Documents

As previously mentioned, the environmental documents must address human, biological, and physical resources. Specific issues associated with each are discussed in the following sections.

Human resources. Issues that could affect the human environment are land use, employment, housing, educational opportunities, transportation, local economics, regional economics, recreation, historic resources, cultural resources, public services, noise, visual, aesthetics, and social factors. Identifying direct effects of watershed programs on human resources can be determined by reviewing general land-use plans and recreational use data, comparing population and economic demographic data and projections, reviewing the site, and reviewing previously identified historical and cultural resources. The results of the review process are compared with the proposed effects of the alternatives.

The difficulty in evaluating effects to human resources results from the interrelated nature of the issues. For example, a new dam may increase flat-water recreational opportunities but reduce whitewater rafting or upper watershed fishing opportunities. The changes in recreational opportunities would cause a direct change in recreational-related employment. These changes may affect regional income, which in turn may lead to changes in affordable housing. In addition, the quality of recreational activities could affect lifestyles and the responses to these changes, such as the need to travel farther to participate in specific recreational activities.

One of the most difficult steps in environmental planning is to determine the significance of the effects. The overall change in any one area may not be quantitatively high, but the effect to specific community sectors may be very significant. Therefore, presenting the information in a manner that will reflect the significance of the effect is important. For example, the loss of 200 jobs may not be significant in a county with more than 1,000,000 people. However, if the jobs are located in a town with a population of fewer than 5,000, the economic and social effects would be very high. This type of information must be presented to the lead agency in a fair and appropriate manner, and in a way that allows the affected community members to participate in the planning process. An interactive planning process will also allow the utility to define acceptable mitigation measures to reduce the effects of potential alternatives.

Biological resources. Biological resources include fish, wildlife, and vegetation. The descriptions of the current conditions and the potential effects are usually segregated into these three broad categories. Further categorization occurs based on the potential effects of permitting or approval agencies' proposed actions and requirements. Surveys are usually completed as part of the environmental planning process. The level of effort in the survey ranges from a review of literature, including information compiled by resource agencies, to detailed delineation surveys. The US Fish and Wildlife Service (USFWS) and similar state agencies have compiled a significant amount of data concerning habitat types and the presence of habitats that may support special status species, including listed or candidate species that may be threatened or endangered. Results from the literature review may be used to screen preliminary alternatives and focus future survey efforts.

Information evaluated in the environmental planning process should be sufficient to identify adverse effects or benefits on all life stages of fish, wildlife, and vegetation. Therefore, the analysis must consider resident and migrating fish and wildlife species. For special status species, conducting field studies to determine the presence of species or their habitats may be necessary. If the species migrate or can be viewed only during a specific time period, scheduling the fieldwork becomes part of the critical path. This may occur for migrating species or for species with long hibernation or estivation cycles. For example, some beetles can be viewed for only a four-to-six-week period during the flowering of specific plants because they hibernate the rest of the time.

The analysis of biological resources must also consider the interrelationship of fish, wildlife, and vegetation. If vegetation is restored along a stream, it could reduce water temperatures and provide more food that would increase fish populations. However, the increased vegetation would also provide more cover for foraging wildlife species that may change the balance of predation for fish and wildlife species. This type of change is appropriate in a balanced ecosystem. However, if the ecosystem does not include adequate species diversity, the increased predation could lead to an increased number of species that may be pests for adjacent land uses.

If the habitat is conducive to special status species, the biologists should initiate a consultation immediately with the USFWS for federal projects or projects with federal permits, or with the state resource agency. These types of surveys must provide information in accordance with Section 7 of the Endangered Species Act (ESA) or similar state laws. For actions subject to Section 7 of the ESA, the USFWS must complete a consultation to assess potential effects on the species. If there are potential effects, mitigation plans will be required to avoid any activity that could lead to the death of an individual or an entire species. If the effects cannot be fully mitigated, the USFWS completes a biological assessment to determine whether the actions could jeopardize the species or seriously alter its habitat. If a "Jeopardy Opinion" is made, a "Biological Opinion" is developed to identify reasonable and prudent alternatives to avoid jeopardy.

Historically, options considered to avoid jeopardy have been site specific and frequently would allow only the "No Action" or "No Project" alternative. However, recently, preservation and enhancement programs for selected areas of considerable size or quality have been implemented in several locations throughout the country under a "Habitat Conservation Plan." These programs are generally located adjacent to the site of the proposed project, but the areas may be in a more appropriate portion of the watershed. Through this program, special habitat banking can occur to provide a large area that will support a healthy ecosystem. This has been more successful in restoring populations than in maintaining small areas of undisturbed land within urban or agricultural environments.

Physical resources. Issues that could affect the physical environment range from surface water, groundwater, air quality, soils including hazardous materials from native materials or waste products, geological risk caused by seismic or land instability, and energy resources. Information concerning these resources can be obtained from field investigations, numerical modeling, literature review, or qualitative analyses. If the information is obtained from numerical modeling, the models can also be used to evaluate the alternatives.

The evaluation of physical resources is interrelated with biological and human resources. For example, surface water affects fishery and wildlife, groundwater, local sector economics, recreation, and other resources. Therefore, surface water resources may need to be evaluated for changes in flow quantity and quality before initiating other resource analyses.

Permits and Approvals in the Environmental Process

Environmental planning documents are used for two reasons. First, as previously discussed, decision makers and the public use planning documents to understand potential effects of the alternatives. Second, the documents are used to obtain permits and approvals for constructing and operating facilities. State and local agencies grant similar permits and approvals. Sometimes, state or local agencies will accept the documentation provided to federal agencies; however, they often require utilities to submit separate documentation. Some of the key federal permits and approvals that must be considered for projects with US federal interests include those related to the following acts.

- *National Environmental Policy Act (NEPA).* This act provides a commitment that federal agencies will consider the environmental effects of their actions. It also requires that an environmental impact document be included in every recommendation or report on proposals for legislation and other major federal actions that significantly affect the quality of the human environment. The environmental impact document must provide detailed information regarding the "No Action Alternative" and "No Project Alternative," the environmental effects of the alternatives, potential mitigation measures, and unavoidable adverse environmental effects. The legislation and associated guidelines also specify public involvement and review requirements.
- *Endangered Species Act (ESA) and Critical Habitat.* The ESA established a national program for conserving threatened and endangered species of fish, wildlife, and plants and for preserving the ecosystems on which they depend. Section 7(a) of the ESA requires federal agencies to consult with the USFWS and/or the National Marine Fisheries Service on activities that may affect any species listed as threatened or endangered. Environmental impact documents must analyze the effects of various alternatives on listed species. These potential effects require utilities to initiate the Section 7 consultation process.
- *Fish and Wildlife Coordination Act (FWCA).* The FWCA requires that parties consult with the USFWS when any water body is impounded, diverted, controlled, or modified for any purpose. The USFWS and state agencies charged with administering wildlife resources conduct surveys and investigations to determine the potential damage to wildlife and to develop mitigation measures. The USFWS incorporates the concerns and findings of the state agencies and other federal agencies into a report that addresses fish and wildlife concerns and recommends ways to mitigate effects caused by federal projects. Compliance with the FWCA is coordinated in conjunction with the ESA.
- *National Wildlife Refuge System Administration Act.* This act consolidates various categories of wildlife ranges and refuges for management under one program. The act protects wildlife and refuge lands from destruction and injury. It also authorizes the regulation of hunting and fishing within refuge boundaries. If a

watershed includes refuges, the environmental planning document should consider the effects of changing adjacent land uses on the refuges.

- *Migratory Bird Treaty Act.* This act requires that lands, waters, or interests acquired or reserved for purposes established under the act be administered under regulations promulgated by the Secretary of the Interior. These regulations must conserve and protect migratory birds in accordance with treaties between the United States and Mexico, Canada, Japan, and Russia. Furthermore, these regulations must protect other wildlife, including threatened or endangered species, and must restore or develop adequate wildlife habitats. The migratory birds protected under this act are specified in the respective treaties. In regulating these areas, the Secretary of the Interior is authorized to manage timber, range, agricultural crops, and animal species, and to enter into agreements with public and private entities. If the watershed is located on a major waterfowl flyway in the United States, the environmental planning document should identify and describe effects to the life stages of the waterfowl that use the watershed.
- *Clean Water Act (CWA).* The CWA restores and maintains biological, chemical, and physical qualities of the nation's water. It also established the National Pollutant Discharge Elimination System to reduce pollutant loads into the nation's water from all point and nonpoint discharges, including stormwater runoff. It provides for development of basin management plans to identify sources of pollution and options to reduce the mass loads. It also requires states or primary agencies to prepare plans to protect wetlands.

  Watershed programs must consider several CWA provisions. Most must consider the effects of nonpoint source runoff, including contributions from undeveloped land, such as the forests that may contribute sediment loading. Implementing new programs in a watershed may require the mass loading from the entire watershed to be evaluated to reduce future pollutant concentrations. Recently, watershed programs have been implemented to identify total maximum daily load contributions for critical constituents. Under these programs, one or more contributors of pollutants may coordinate with other dischargers to reduce the pollutant load in the most cost-effective manner. For example, in a watershed with urban and agricultural users, if an urban user plans to increase organics loading for growth, the most cost-effective alternative may be to fund a sediment removal system for the agricultural area rather than provide a high level of treatment in the urban area.

  New water resource facilities may require construction methods that include dredging or filling of waters or wetlands. In these cases, CWA provides regulatory authority to the US Army Corps of Engineers (Corps) to regulate these activities through a Section 404 permit program. This program integrates with Section 401 of the CWA, which requires that all such permits comply with the overall water quality plan for each watershed. Section 404 permits require alternatives to be implemented that do not require dredging or filling if any practicable alternative does not have more adverse effects than a proposed action. The Section 404 permit process should be initiated early in the planning process and modified if the planning process is changed.
- *The Rivers and Harbors Act of 1899.* This act prohibits the unauthorized obstruction or alteration of navigable waters without approval of the Corps. This permit process is usually combined with Section 404 of the CWA process.
- *Section 106 of the National Historic Preservation Act (NHPA).* This section of the NHPA requires that federal agencies evaluate the effects of federal undertakings on historical, archaeological, and cultural resources, and afford the Advisory

Council on Historic Preservation opportunities to comment on the proposed undertakings. The first step is to identify cultural resources included (or eligible for inclusion) on the National Register of Historic Places that are located in or near the project area. The second step is to identify the possible effects of proposed actions. The lead agency must examine whether feasible alternatives would avoid such effects. If an effect cannot reasonably be avoided, measures must be taken to minimize or mitigate potential adverse effects. Compliance with Section 106 is important, but it is also important to keep it in context with the following three requirements: (1) it does not require preservation of all historic properties; (2) it does not require accurate restoration of historic structures, because accurate restoration may not be possible, economically feasible, or desirable; and (3) it does not prevent new construction in historic areas.

- *American Indian Religious Freedom Act (AIRFA).* AIRFA protects Native American inherent rights to the free exercise of their traditional religions. This has a broad scope and requires consultation with federally recognized tribes when project actions could affect the practice of traditional religions.
- *Archaeological Resources Protection Act.* This act protects archaeological sites on public and Indian lands, establishes permit requirements for excavating or removing cultural properties from public or Indian lands, and establishes civil and criminal penalties for the unauthorized appropriation, alteration, exchange, or other handling of cultural properties. As part of the cultural resources analysis in an environmental planning document, the potential for archaeological sites must be evaluated. However, disclosure of specific sites or specific information about sites may not be allowed because of the need to preserve the sites from scavengers and tourists. Therefore, the environmental planning document frequently includes a separately bound appendix with specific archaeological information that may be summarized for the lead agency.
- *Indian trust assets (ITAs).* The US government's trust responsibility for Indian resources requires federal agencies to protect and maintain trust resources. These responsibilities include taking reasonable actions to preserve and restore tribal resources. ITAs are legal interests in property and rights held in trust by the United States for Indian tribes or individuals. Indian reservations, rancherias, and allotments are common ITAs. All federal environmental planning documents must consider the presence and potential effects of alternatives on ITAs in the study area.
- *Indian sacred sites on federal land* (Executive Order 13007). This order provides that in managing federal lands, each federal agency with statutory or administrative responsibility for managing federal lands will, to the extent practicable and as permitted by law, accommodate access to, and ceremonial use of, Indian sacred sites by Indian religious practitioners and avoid adversely affecting the physical integrity of such sites.
- *Environmental justice (EJ).* EJ pursues equal justice and equal protection for all people under environmental statutes and regulations. EJ helps ensure that no one socioeconomic group is asked to bear more than its share of adverse environmental impacts. In the past, low-income, minority, and/or Native American communities may have borne more than their share of such effects. Under NEPA, the NHPA, the ESA, and others, it is important to identify and address EJ concerns. EJ requires federal agencies to analyze environmental and health effects, and the economic and social effects of federal actions. It also requires mitigation measures to be identified. The importance of public participation is emphasized so that

affected communities can provide input to the regulatory agencies (see chapter 2 on public participation).

- *Managing floodplains* (Executive Order 11988). This order considers alternatives to avoid adverse effects in the floodplain or to minimize potential harm. In addition, the order requires federal agencies to evaluate the potential effects of any actions they might take in a floodplain and to ensure that planning, programs, and budget requests reflect consideration of flood hazards and floodplain management.
- *Protecting wetlands* (Executive Order 11990). This order authorizes federal agencies to take actions to minimize the destruction, loss, or degradation of wetlands, and to preserve and enhance the natural and beneficial values of wetlands when undertaking federal activities and programs. Any agency considering a proposal that might affect wetlands must evaluate factors affecting wetland quality and survival. These factors should include the proposal's effects on the public health, safety, and welfare caused by modifications in water supply and water quality; maintenance of natural ecosystems and conservation of flora and fauna; and other recreational, scientific, and cultural uses.
- *Wild and Scenic Rivers Act.* This act designates qualifying free-flowing river segments as wild, scenic, or recreational. The act establishes requirements applicable to water resource projects affecting wild, scenic, or recreational rivers within the National Wild and Scenic Rivers System, as well as rivers designated on the National Rivers Inventory. Under the act, a federal agency may not help construct water resources projects that would have direct and adverse effects on the free-flowing, scenic, and natural values of a wild or scenic river. If the projects would affect the free-flowing characteristics of a designated river or unreasonably diminish the scenic, recreational, and/or fish and wildlife values in the area, such activities should be developed in consultation with the National Park Service and in a manner that minimizes adverse effects.
- *Federal Water Project Recreation Act.* This act requires parties that are planning any federal navigation, flood control, reclamation, or water resource project to fully consider opportunities that the project affords for outdoor recreation and to enhance fish and wildlife. The act requires planning with respect to developing recreation potential. Projects must be constructed, maintained, and operated to provide recreational opportunities consistent with the purpose of the project.

Key Aspects of Environmental Assessment

The evaluator may want to consider the following during the assessment of environmental aspects of various sources of supply:

- *Rivers.* Rivers generally tend to be dynamic and biologically diverse compared with other water supply alternatives. To better protect aquatic life and accommodate downstream competing uses, instream flow requirements have been steadily increasing. These restrictions can significantly reduce both the viability and available yield of direct river withdrawals, onstream reservoirs and pumped-diversion reservoirs.
- *Reservoirs.* Main-stem (on stream) reservoirs many times will have a greater level of overall environmental impact than reservoirs on smaller tributary streams, because of typically increased wetland and habitat disturbance, impediments to fisheries resources, disruption to free-flowing streams (stream bifurcation), and other factors. US Geological Survey maps and other pertinent mapping should be investigated to initially define extent of potential effects due to wetlands, protected

lands, natural gas pipelines, electrical transmission lines, historic sites, relocation of residents, pollutant sources, and so on. Subsequent site visits may be warranted for field verification. Instream flow restrictions can affect the viability of onstream and pumped reservoirs. Expansion of existing reservoirs by raising the water level may be attractive because of cost and environmental considerations.

- *Desalination.* Concentrate waste can typically be characterized by low dissolved oxygen, low pH, and may contain corrosion products and residual chemicals from pretreatment processes. Concentrate is toxic to marine and freshwater organisms, which makes surface disposal very difficult. Low recovery systems also create a large volume of concentrate that is also hard to dispose. Six common methods for waste concentrate disposal are surface water discharge, indirect ocean discharge from a regional concentrate interceptor to a wastewater treatment plant not producing reclaimed water, land application, injection well, evaporation, and direct ocean discharge through an outfall.
- *Demand management and water conservation.* Reductions in water use can reduce water withdrawals from supply sources resulting in higher instream flows for fish and wildlife habitat.
- *Water marketing.* A major concern associated with any water transfer or exchange is a potential adverse effect on the environment. Water transfers can degrade surface and groundwater quality, affect aquatic and terrestrial flora and fauna, reduce recreational water uses, cause aesthetic problems, and subside the ground surface, affecting public infrastructure (such as drainage systems and flood control levees in the area of excessive groundwater pumping). Effects can be local or regional, depending on the size and length of the transfer and the prior base environmental conditions. In general, short-term and intrabasin transfers have relatively minor potential environmental impacts and are therefore subjected to much less environmental scrutiny than long-term and interbasin transfers. Environmental analyses required to facilitate transfers varies among the states and is highly dependent on the environmental compliance requirements of the agencies involved.

When comparing a wide range of alternative sources, it may be necessary to retain alternatives that appear less likely to make a short list until further analyses required for permitting are complete (e.g., an endangered species was found in the reservoir area).

IMPLEMENTATION ASSESSMENT

Assessing implementation aspects when comparing a list of alternatives reflects the ease or feasibility of bringing online new sources of supply. Specifically, this criterion measures the

- ease of obtaining federal, state, and local permits;
- effects on tribal treaty interests;
- public and political support;
- institutional barriers;
- project timing and readiness;
- ability to spread costs over time;
- ability to expand beyond the planning period; and
- ability to accommodate entrance or exit of members.

Implementation Considerations

Implementation can, in some cases, be the most important criterion in the alternatives selection process because it addresses the element of timing for bringing new supplies online. A summary of key considerations associated with implementation is outlined below.

- *Permitting.* The time frame for conducting the technical and environmental studies needed to support the permits for major or complicated water resources projects can be 5, 10, or more years in many cases. Adequate time must be included for the studies, regulatory reviews, meetings, public notices, and so on.
- *Public and political support.* Water supply projects that can serve as regional facilities are generally viewed favorably by politicians and regulators. The evaluator should carefully consider the anticipated public and political response to a given alternative. Relevant issues include overall citizen support for, or opposition to, the project; enhancement of local or regional quality of life or recreational value; regional or statewide support; financial stability; tax and water rate implications; and general environmental soundness.
- *Project timing and readiness.* Frequently, grants and matching funds from states and the federal government have to be encumbered within a given period of time, so projects sometimes must be "shovel-ready" (i.e., ready to start) to be eligible.
- *Ability to spread costs over time.* A distinct advantage associated with certain water supply alternatives is their ability to spread costs over time and avoid a one-time capital expenditure. In some cases, staged development can be a cost-effective and financially desirable means of meeting growing water demands. Small increments of additional supply, such as incrementally developing wells in a well field, should be brought online when economical. This prudent approach allows capital costs to be deferred to match the growing revenue stream. If a major, staged construction project is considered, permits should be obtained if possible for the entire project before the first stage of construction. When the initial phase of water supply development is economically attractive by itself, a greater level of risk related to permitting of subsequent phases can be more readily accommodated. However, permits and allocations for the ultimate level of project development should be obtained when possible.
- *Ability to accommodate entrance or exit of members.* When entering into agreements for water supply projects that serve a number of members in the region, it is important that members have the ability to enter and exit the organization over time. As water supplies are fully utilized and infrastructure ages, potential members may decide to decommission its facilities and enter into an agreement with a regional organization.

It is important that the evaluator understands the factors associated with determining project scope and staging its implementation. Investigations, planning, environmental studies, permitting, and other activities required for developing a new source of supply can be rigorous, time-consuming, and costly endeavors that water utilities will not want to repeat often to accommodate stepped expansion of their water systems. Additionally, over the past several decades, permitting requirements have shown a clear trend toward being more rigorous and demanding in terms of both time and cost. This trend is likely to continue for the foreseeable future. Therefore, when the following four conditions are present, a long-term solution should be pursued:

1. Project permitting will be rigorous, costly, and time-consuming.
2. Continued long-term growth beyond the planning period can be demonstrated.
3. An environmentally, technically, and economically viable large-yield project alternative (greater than or equal to the planning period shortfall) has been identified.
4. Financial resources are sufficient to accommodate a larger project step size.

Key Aspects of Implementation Assessment

The evaluator may want to consider the following during the implementation assessment:

- *Rivers.* Alternatives can have long time frames for implementation because of studies and permitting associated with competing uses on larger river systems.
- *Reservoirs.* Typically, there are long time frames for bringing reservoirs online because of extensive technical and environmental studies and permit requirements. Reservoirs can be implemented on a multipurpose basis to garner political and public support, but they are not easily expanded beyond the planning period.
- *Groundwater.* Compared to other alternatives, groundwater systems tend to be easily expanded by bringing new wells online.
- *Demand management.* Aggressive demand management and water conservation can result in cost savings associated with downsizing or deferring capital projects. The effectiveness of programs are frequently dependent on customer participation and incentives.

COST ASSESSMENT

When comparing a list of water supply alternatives based on cost, it is important to keep in mind the following key considerations:

- *Level of detail and uncertainties.* Conducting the studies and investigations necessary to generate sound and defensible cost estimates for making informed water supply decisions can be time-consuming and expensive. The primary objective when evaluating a list of alternatives is to create an "apples to apples" comparison. It is appropriate that as more rigorous investigations are conducted through the course of a study, the level of detail for the components included in the cost estimates increases for those alternatives carried forward. It is also important to have a strong understanding of the uncertainties (i.e., confidence level) associated with each of the components of the cost estimate and convey this information to decision makers.
- *Time period of analyses.* When assessing life-cycle costs, it is important to consider both the time period of analysis and the discount rate. Costs for project alternatives are often distributed differently over time. Some projects require large initial capital expenditures, with relatively small ongoing operating expenditures (reservoirs, transmission mains, etc.). Other projects may require modest initial capital expenditures but have several increments phased in over time and high operating expenditures. Because alternative projects will differ in the combination of initial and incremental capital spending and ongoing operation and maintenance (O&M) expenditures, it is recommended that the analysis be based on the full life cycle of the projects. Developing a cost model that captures both the pattern and timing of project costs may be warranted. An appropriate discount rate should be used to discount all future costs into their present worth to compare costs that occur at

different times in the future. Using information on the magnitude and timing of the capital and O&M costs, a utility can determine how best to adequately fund projects over time. Note that a long period of analysis (e.g., 50 to 75 years) tends to favor longer-lived infrastructure elements (dams, transmission mains, etc.) that are capital intensive and are not likely to require replacement during the time period of the analysis. Conversely, alternatives that are equipment intensive or have high O&M costs tend to benefit from a shorter period of analysis.

- *Quantification.* When comparing alternatives, the evaluator should attempt to shift as many qualitative factors to quantitative values (e.g., costs, or consistent scoring systems) as possible. This is frequently the case for the water quality criteria when comparing finished water quality for surface and groundwater supplies. Both sources may meet the regulatory requirements, but taste and odor or other characteristics may differ significantly. In some cases, it may be possible for the evaluator to incorporate the cost for alternative treatment processes to achieve a comparable level of finished water quality.
- *Perspective.* The evaluator must clearly define the stakeholder perspective (i.e., utility, ratepayer, society), particularly when a number of alternatives with varying characteristics are being considered (reservoirs, desalination, etc.). For a utility, the costs for a water supply include payments the utility makes for construction, engineering and technical services, pilot studies, environmental reviews and permitting, bidding and contracts, administration and legal services, land and rights-of-way acquisition, bond issuance, commissioning costs, construction management, and decommissioning. In addition, any costs associated with O&M (chemicals, operators, utility staff, etc.) need to be included in the cost estimate.

The cost assessment is conducted to evaluate the overall unit cost (e.g., $/mgd) of each alternative under consideration and may include

- capital,
- O&M,
- present worth,
- life-cycle costs,
- treatment or operational functions that rely on natural systems for lower life-cycle operating costs through reduced energy and chemical input, and
- reduced reliance on the energy grid through greater energy efficiency or self-generation of energy.

Cost Considerations

In the following discussion, costs are classified as internal or external to the utility. Internal costs are borne by the entity (utility) that benefits from the project. Internal costs would be included in analyses performed from the utility perspective. Costs borne by an entity that does not receive a benefit from the project are external costs. Capital and O&M costs are internal to the firm or utility that creates the project. These are paid by the utility, directly or indirectly, to achieve a result and are factored into the decision by the utility depending on how it allocates its resources. Many projects, on the other hand, also have the potential to affect parties not directly related to the project. When decisions are made by a utility, these external costs may be overlooked because they must be paid by another entity (e.g., placing restrictions on recreational use of reservoirs). Indeed, from the utility perspective, they are not relevant. For overall economic efficiency, all costs should be accounted for, particularly considering the perspective of the affected community. This requires that the external costs be included to the fullest extent possible.

The following paragraphs describe the various aspects of costs that should be considered when generating cost estimates for water supply alternatives.

Capital costs. Capital costs consist of all expenditures necessary to create a productive asset. They are relatively large expenditures that must be made in a short time. Capital costs are sometimes referred to as "lumpy" expenditures because they must occur during short time frames and cannot easily be incrementally changed. They tend to be all-or-nothing expenditures and must be made before any production can occur.

Financing. Developing a new water resource generally requires substantial resources, usually over a short time, and must be made before benefits (receipt of revenues) are realized. In most cases, the amounts required are well beyond the ability of the utility to fund them through internally generated revenue sources. Consequently, the projects are often financed through external sources, particularly by issuing debt instruments such as bonds. *Debt financing* is the interest and associated costs that must be added to the overall lifetime project cost. Even though the additional interest costs add to project cost, their economic effect will be less because they are distributed over time.

Because costs to the utility are spread over the term of the debt (often 20 to 30 years) rather than entirely up front, a revenue stream can be developed that pays for the debt service (principle and interest). Debt financing may be more equitable or fair because it allows a matching of costs with benefits over time—the benefits of a large project are distributed over an extended time frame; debt financing allows the costs to be distributed over a corresponding time frame. Although debt is spread over many years, it is seldom for the entire life cycle of the project.

Subsidies and grants. Because of the large capital expenditures required for many water resource projects, cost offsets may be available that reduce the internal costs to the utility. Grants or subsidies from other entities, particularly other governmental entities, may be available. To the extent that these can be used, the internal cost to the utility is reduced. If the analysis is performed from the utility perspective, the project cost is reduced by the amount of the grant or subsidy and the project becomes more feasible. If, however, the analysis is performed from the society perspective, the grant or subsidy is a transfer and has no effect on project costs; costs decrease for the utility but increase equivalently to the granting entity.

Operation and maintenance costs. O&M costs represent the ongoing expenditures necessary for the utility to operate the project over time, including keeping facilities in good repair. Unlike capital costs, O&M costs are continuous over the lifetime of the project, depending to some degree on the level of production from the project.

Imposed costs. Most production processes, whether they occur in a factory, office building, or water treatment plant, have the potential to impose costs on someone else. Air pollution from a chemical factory imposes additional costs on those who must breathe the polluted air. Other businesses may have to spend more funds to filter or clean the air to make conditions palatable for their employees or customers. A farmer whose pesticides run off into a river used for drinking water imposes costs on the water utility to remove any residue from the pesticide. These all represent costs imposed by a production process on another entity that has neither consented to the imposition of the cost nor directly benefits from the production process. Imposed costs may be estimated based on the cost of efforts by the other entities to undo their effects.

For the most efficient allocation of resources, costs should be internalized to the extent possible. Where external costs are not factored into the decision-making process, the product tends to be underpriced and overproduced. Because prices are determined, at least in part, by the costs to produce a product, they will be lower than they should be. Because prices are lower than they should be, customers will desire more of the product and overproduction occurs. On the other hand, for production processes that have had costs imposed on them, production costs (to remove pesticides from drinking water, for example) and prices charged tend to be higher than they would be otherwise, leading to a tendency to underproduce.

Environmental costs. Some external costs are imposed not on a specific entity but rather on the environment. Where imposed costs can be estimated and quantified based on the additional costs, other producers must pay to offset the imposed costs; environmental costs are not as easily determined or valued. Valuation of the loss of habitat or of an endangered species is nearly impossible.

Because valuation of environmental costs is very difficult, the costs are often not included in economic analyses, even though they may be recognized in other ways. Integrated resource planning efforts commonly screen alternatives for environmental impacts before performing the economic analysis. Only projects already determined to have acceptable environmental impacts are typically considered in latter phases of the study. Projects deemed likely to have greater environmental impacts than others might have a 10 percent cost applied to account for at least some of the real costs.

Project Benefits Associated With Costs

Identifying and quantifying the benefits of water-related projects can be difficult. Some benefits, such as additional revenues generated by sales, are fairly straightforward, but most are more troublesome. Many economic analyses for water projects concentrate on the cost side only, assuming the benefits are adequate to make all projects economically feasible, and the main task is to identify the lowest cost option to achieve the positive benefits.

Value of water production. The most direct measurement of benefits for water resource projects is the value of additional water sales generated as a result. Because both the price and the quantity of additional water can be projected, valuation of the benefits, at least from the utility perspective, is straightforward. Other sources of benefits from water projects could include power generation and sales, recreation use fees, additional impact fees from growth supported by a new water supply, and additional tax revenues from growth in the service area. All are related directly to increased production and are relatively easy to quantify.

Avoided costs. Many benefits of a project may be indirect, coming about as a result of not having to use resources in some other way. For example, a surface water supply project that replaces a groundwater supply that is seriously overdrafted and is creating saltwater intrusion provides benefits that cannot be measured in additional water sales. Rather, the benefits come from avoiding the costs associated with the saltwater intrusion. If adding a surface water supply can avoid the need to build a desalination facility, the benefit of the surface supply is equivalent to the avoided cost. These avoided cost-benefits can come about as the result of avoiding capital and O&M expenditures.

External benefits. Water resource projects may provide positive external benefits that should be included in determining total benefits, just as the external costs should be included in total costs. For example, creating a lake for water supply purposes may also provide recreation benefits, habitat for wildlife, or flood control benefits. As with external costs, the allocation of resources can be distorted by failure to include these benefits in the economic analysis.

Key Aspects of Cost Assessment

The evaluator may want to consider the following key aspects during the cost assessment for the various sources of supply:

- *Onstream reservoirs.* These facilities offer the benefit of low operating cost compared to pumped reservoirs, because reservoir recharge is derived from natural runoff. When located in upland areas, pumped transmission to treatment facilities can sometimes be minimized or eliminated with delivery by gravity pipe flow or by stream channel releases with downstream diversion at or near the water plant site. Onstream reservoirs are also more likely to require a large and expensive dam and spillway, and the valley areas of larger streams are more likely to have infrastructure development that would be affected by a reservoir (roads, bridges, etc.). In addition, the cost for mitigating environmental impacts is likely to be significantly higher for dams located in valley areas with larger streams or rivers.
- *Pumped reservoirs.* Pumped-diversion reservoirs are somewhat more capital and operational intensive because of the need for, and operation of, pumped-diversion facilities (intake, pumps, and additional pipeline construction). Additionally, many streams of sufficient drainage area for water supply purposes lack the appropriate topography and geology for developing a cost-effective dam and reservoir site for drought storage. Pumped reservoirs typically have less expensive dam and spillway structures compared to onstream reservoirs. The most attractive aspect of pumped reservoirs is that the effects to the environment are minimal compared to dams located in large valleys.
- *Groundwater.* Drilling a pilot hole for large wells should be employed to collect data (i.e., water levels, water quality samples, aquifer properties, etc.) and minimize cost. After the analysis of the pilot hole data is complete, the pilot well can be abandoned or converted to a production well. The advantages of using groundwater in comparison with other sources is the ability to locate wells near demand centers to limit infrastructure piping, and it often does not require robust treatment and typically has minimal environmental impacts.
- *Aquifer storage and recovery.* In addition to the water treatment plant costs, the evaluation of ASR projects should consider the costs for installation and operation of the necessary piping to convey treated water to the injection wells, for the injection well pumps themselves and for the production wells to return the recovered water back to the point of use. Because all of the water to be pumped is of potable quality, the transmission lines can carry both the water to be injected and the water to be recovered in the associated opposite directions.
- *Infiltration galleries.* When infiltration galleries are located below or adjacent to a water body, they tend to have higher turbidity and higher bacteria count because of less extensive filtration. Where such withdrawals are considered to directly affect the quantity or quality of surface flows, many of the considerations presented for direct surface water withdrawals will need to be followed, including meeting Surface Water Treatment Rule requirements. This water would likely require filtration, which increases the cost of infiltration galleries.
- *Reclaimed water.* Because relatively large storage volumes are needed for relatively small-capacity reclaimed water systems, seasonal storage is most often constructed in open-storage reservoirs. Because these usually require relatively large land areas, a particular range of elevations, a dam, and other appurtenances, such reservoirs are often not constructed because of considerable expense and environmental impact. In addition, the capital and O&M costs for treatment appropriate for the intended end use and the ability to meet strict water quality standards needs to be considered.
- *Desalination.* Concentrate disposal costs can be a substantial part of the total cost of desalting and must be considered in the initial planning and design. When using

either membrane or thermal separation processes, highly concentrated reject water (concentrate) is produced along with high-quality potable permeate. The total dissolved solids of the concentrate depend on the source water characteristics. Many regions of the country employing desalination are also providing alternative sources of energy to offset treatment costs.

- *Water conservation.* Water conservation can reduce short-term and long-term O&M costs because of lower energy expenses associated with decreasing need for treatment chemicals and pumping for distribution. In addition, aggressive water conservation can result in cost savings associated with downsizing or deferring capital projects. The two basic disadvantages—lower revenues and concerns over drought conditions—can be addressed through preventive measures, such as periodic rate adjustments and developing a drought contingency plan.

SUMMARY

The goal of the water supply assessment is to generate a matrix summarizing the characteristics for each alternative based on a defined set of evaluation criteria. This evaluation matrix can then be used by the decision makers to select an alternative, or portfolio of alternatives, for implementation (discussed in chapter 9). By conducting the water supply assessment in phases, the evaluator can focus on those alternatives that are most promising, thereby potentially achieving significant cost savings by avoiding or minimizing the need for costly technical and environmental studies.

Chapter 9

Alternatives Analysis

Water resources planning inevitably involves comparing a variety of possible future courses of action. These could include whether to expand the community's water supply portfolio by tapping into a new surface or groundwater source, adding or increasing the level of water reuse, examining desalination, increasing water efficiency programs, expanding reservoir storage capacity, or any of the other alternatives described in chapter 4.

Such planning also entails asking and answering questions about (1) why any action may be warranted (as contrasted to a "do nothing," status quo baseline), (2) determining what option will best serve the community, (3) when action should be initiated, and (4) how to best implement a selected option. As such, water resources planning entails evaluating the various alternatives that are feasible for a utility and making informed decisions about which option may be considered the "best" choice.

This chapter examines the issues and methods associated with conducting (and communicating) a sound comparison of alternatives. In addition, various established methods for comparing alternatives are also described.

A common theme across the various evaluation methods is that any analysis of alternatives is based on evaluating how each potential water supply option performs according to several key criteria. These criteria generally include financial cost, environmental factors, community and related social values, and other key factors (as described in chapter 8). The ultimate objective of the analysis is to provide useful and objective information to decision makers, so that they can identify and support the most sustainable and suitable alternative.

BASIC OBJECTIVES AND GUIDING PRINCIPLES

It is always good practice to make a sound and effective business case—or value proposition—when a utility is seeking support for a large capital outlay or other investment, such as is involved in water resources planning. Utility managers, governing boards, and economic regulators (such as public service commissions) are keenly interested in moderating future rate increases on water utility customers, especially in difficult economic times. They need to be thoroughly convinced that there is a real problem, that there are high costs or other serious consequences if the utility fails to act, and that the proposed investment (i.e., the selected alternative) is a wise choice that will leave the community better off than if no action were taken (and also better off than if an alternative potential solution were selected).

Utility customers and local stakeholder organizations may also be skeptical of the need for a specific large-scale water resource investment that their utility is considering. Concerns may reflect a broader array of issues than just the cost and the ultimate effect on rates. There may be strongly held opinions about the environmental or social consequences of a proposed utility project. For example, utilities often face opposition to efforts to invest in desalination, water reuse, reservoir expansion, or other such options to increase local water supply reliability for the community as it grows and as its current water sources become fully tapped. Such public concerns extend beyond the overall fiscal costs and often include potential effects on local ecosystems, community identity, affordability, energy demands, carbon footprints, and so forth.

Ultimately, utility professionals need to be able to convince themselves, their managers and boards, their customers, and other stakeholders that they are proposing a wise course of action. They need to effectively communicate that even if a proposed action entails a costly investment, the community will be better off for having taken that action.

Principles that should guide a utility's efforts in conducting a sound analysis of alternatives include objectivity, transparency, and replicability. These are generally widely recognized elements of good practice. It also is important to be as comprehensive as possible (i.e., avoiding omission of important benefits or costs, avoiding double counting, discounting to present value) and include qualitative as well as quantitative and monetized outcomes.

WHY CONDUCT AN ANALYSIS OF ALTERNATIVES?

Tools are needed to help water agencies and other relevant entities and stakeholders understand how to assess and portray the overall advantages and disadvantages (e.g., benefits and costs) of the water resources planning and management options that are available to them. As stewards of water resources and physical assets, the environment, and public trust, water agencies need to consider the full range of benefits and costs associated with their activities. This means that they need to take into account all of the effects—good, bad, and uncertain—that their activities may impose on the community.

An evaluation of water supply alternatives can be approached using a wide range of tools and methods that are ultimately all related to each other and are variations of the same basic notion—that of making a "business case" or "value proposition" for why one water supply option (including the possibility of not adding a new supply to the portfolio) makes more sense than the other alternatives. These various approaches may be called business case evaluations (BCEs) at a generic level and often consist of some form of benefit–cost analysis (BCA), triple bottom line (TBL) assessment, or other related approach. Each method is described briefly in this chapter. In addition, cost-effectiveness (C-E) analysis, multiple-criteria decision analysis (MCDA), and life-cycle cost analysis (LCCA) are variations on the same theme and are described, compared, and contrasted with the other approaches.

The Challenges Facing Utilities

Water agencies, regulatory bodies, the research community, and other entities have identified and implemented a suite of practices, processes, and regulations that support the safe and prudent management of water supplies. Still, numerous significant challenges face water agencies and other interested parties as they consider how to best manage their water supply portfolios and provide suitable stewardship of their associated source and receiving waters. As introduced in chapter 8, these challenges reflect practical realities, including the following factors:

- Water supply planning and related water resource management options tend to require high capital outlays and operating expenses (including energy costs). Managers need to consider whether the benefits warrant the high costs of some of the options available to them.
- Many water supply and resource management options tend to be controversial because of public concern over various real or perceived risks and inconveniences. These potential negative effects may include suspected and actual risks to environmental resources, climate change, and/or public health. Managers need to consider a broad suite of water resource options and weigh the potential outcomes associated with each of those options.
- Physical or other constraints often limit the technical feasibility, constrain the institutional viability, or affect the cost of some water supply options for some locations. For example, some water rights laws may limit water reuse opportunities, or the lack of suitable hydrogeologic conditions may make aquifer storage and recovery infeasible. Therefore, benefit–cost type comparisons are often highly site and utility specific, and comparing the options can be quite complex.
- Cost should not be the only consideration when making decisions, because costs tend to be short term versus long term. Long-term solutions may be ignored for short-term savings. Furthermore, water supply alternatives are often associated with different types and levels of benefits, which implies that basing a decision solely on a least-cost criterion may lead to selecting an alternative that does not maximize the net value to the utility and the community it serves.

BCA and related economics-based decision support tools—if conducted with a suitably broad "social" perspective and based on a transparent application of appropriate methods and data—offer a well-established framework for addressing many of these issues. Water agencies need a practical way to better recognize and account for the effects of the water resource alternatives they are considering, regardless of what type of effect, who realizes the benefits, or who bears the costs and risks.

It is also important to recognize that conducting BCAs and related BCEs is often a complex and specialized endeavor. In some cases, utilities may need specialized technical expertise to assist in the framing, execution, and/or presentation of these analyses.

Recognizing That BCA Is Simply One Tool in the Decision Support Toolbox

Although BCE approaches such as BCAs and TBL assessments are valuable tools, it is important at the outset to acknowledge their inherent limitations:

1. BCAs and related approaches for analyzing alternatives are but one of several tools that water agency managers and public officials need to have in their decision support toolbox to effectively deliberate water supply options and decisions in the public policy sphere. While BCAs can be very informative and useful, they do not necessarily supplant other tools or override other considerations.
2. BCAs and related approaches are not an exact science and should not be seen as providing a firm "rule" to determine what outcomes (e.g., supply alternative) should be pursued. Rather, BCA and related methods are simply "tools" to help systematically organize information and illustrate suitable comparisons across options. Analyses of alternatives are tools intended to inform decision makers and

communicate information to stakeholders; they are not rules about what decision should be made.

3. A good BCE alone is unlikely to carry the day for a manager or public official attempting to address water supply planning and management options. Instead, a BCE is among a suite of tools and can be used to complement other types of analyses, perspectives, and communication approaches.

Furthermore, BCA is but one approach drawn from a suite of related BCE-type economic evaluation approaches that can be adapted to analyzing options. For example, the TBL assessment is a planning tool developed to help agencies track their progress toward promoting sustainability. There are many variants of TBL analysis that have been applied. TBL analysis is often used as a streamlined and largely qualitative version of a broad social accounting-oriented BCA. However, a highly *quantitative* form of TBL analysis can also be developed as a robust form of a BCA in which all the effects are allocated across the three "bottom lines," reflecting financial, social, and environmental outcomes.

OVERVIEW OF BUSINESS CASE EVALUATION METHODS

As noted previously, there are various approaches available for evaluating alternatives. First, a distinction is made between financial analysis and economic analysis. Both have important roles in the evaluation process. Second, an overview is provided of various tools and approaches to alternatives analysis, so that their commonalities and differences can be more clearly appreciated by utility managers and other practitioners.

Distinguishing Financial Analysis From Economic Analysis

An important distinction is the difference between *financial* analysis and *economic* analysis. This chapter focuses on the latter, emphasizing variations of benefit–cost comparisons as the applicable form of economic analysis.

To make the distinction more clear, a *financial* analysis of water supply management focuses solely on the *cash flows* of expenses paid out and revenues brought in to the utility. In other words, a financial analysis focuses on the internal monetary bottom line of the agency and disregards any effects or values that do not register internally within the utility as cash transactions.

This means that a financial analysis will reveal, for example, how much revenue is collected and how much money the utility spends (salaries and other operating expenses, and debt service on capital equipment investments).

This comparison of cash outlays (expenses) to cash inflows (revenues) produces the utility's *financial bottom line*—the degree to which cash expenses may exceed inflowing revenues (or vice versa). If the utility were a private sector enterprise, this would indicate whether or not the firm earned a profit (revenues exceeding costs) or a loss (costs exceeding revenues) for a given service or product line.

In contrast, an *economic* analysis may start with the financial analysis, but then adds external and nonmarket effects to provide a much broader perspective of the full value (benefits) and full costs of the water planning and service provision operations.

In other words, an economic analysis provides a suitable benefit–cost perspective for considering if a water resource option is an investment that is worth the expense, given due consideration to all the values provided to (as well as costs imposed on) the broader region and community as a whole.

The comparison of all the benefits generated (regardless to whom they accrue) to all the costs (including any costs borne outside the utility) indicates whether there are positive net benefits (i.e., benefits that outweigh costs) for society as a whole. In essence, this entails including all notable "externalities" into the assessment.

Adding Value to the Equation

Financial analyses are very important and essential in many ways. Nonetheless, they typically provide too limited a context with which to evaluate the total social worth of most water projects. The cash flows evaluated in the financial analysis are not the same as the true worth or *value* of most water resource projects to the community and society as a whole.

In economic parlance, an alternatives analysis should entail conducting a *full social cost accounting* of the benefits and costs of projects. This approach can be used to estimate and portray environmental and other costs and benefits of a water option relative to other alternatives in an objective and comprehensive manner, and show how the benefits of an alternative compare to its costs.

Including Internal and External Values for Market and Nonmarket Goods and Services

Benefit–cost analysis and associated variations are intended to reflect an objective and empirically sound effort to describe, quantify, and monetize to the extent feasible all of the outcomes associated with an alternative, regardless of who bears the brunt of the effect or whether the effect can be valued using observed market prices.

The preferred approach is to try to include benefits, costs, and risks borne "internally" by the water agency as well as those effects borne "externally" by other parties (households, businesses, plant and field workers, etc.). The approach also strives to include, to the greatest degree feasible, "nonmarket" goods and services—meaning effects that are not typically traded in markets and therefore do not have market-observable values (e.g., for ecosystem services provided by instream flows, which support aquatic species and may in turn support a local recreational sector of the economy). These values may be estimated using nonmarket valuation techniques, which are described in greater detail later in this chapter.

Overview of Various Evaluation Approaches

Benefit–cost analysis (BCA) is intended to reflect the broad range of outcomes and effects of an alternative. The typical output is to consider which option provides the greatest "net social benefit," which means the option where the overall benefits exceed costs by the greatest amount (this is also where marginal costs equal marginal benefits). Often mischaracterized as including only monetary measures, a good BCA should be comprehensive and include all outcomes, even if some cannot be readily quantified and/or monetized. This means that some important benefits or costs may need to be described in qualitative terms, and it is then up to the analyst to help reviewers understand how the qualitative outcomes may or may not sway the outcome relative to the monetized results. BCA is a well-known approach, and there are many references that can provide useful practical guidance and examples. There are challenges associated with estimating values for nonmarket outcomes, but there also are good resources and tools available to assist in this, as described later in this chapter.

Triple bottom line (TBL) assessment is an approach that can be applied in many forms, ranging from highly descriptive and subjective to more objective and quantified outcomes. As implied by the name, the effects are organized and portrayed according to three bottom lines:

1. *Financial* reflects the cash flow implications of an alternative, such as revenues gained and expenditures or other costs incurred.
2. *Social* embodies the effects of the alternative on the broader community, such as public health and welfare, water system reliability, contributions to employment or other community values, affordability, and so forth.
3. *Environmental* captures the effects of the alternative to watersheds and other ecosystems, carbon footprints, and other consequences for natural systems.

The TBL concept emerged from the sustainability field as a suggested approach for corporations and other entities to expand how they conduct their annual reporting (Elkington 1998). In lieu of reporting a single financially oriented, accounting-based bottom line in an annual report, the suggestion was made to have businesses and public sector entities report annually on how their activities also affected social and environmental matters. Water utilities in Australia now provide such annual, enterprise-level TBL reporting on a routine basis.

Cost-effectiveness (C-E) analysis is typically a simpler, partial form of BCA in which the costs of options are compared when those options essentially produce the same type and level of desired outcome. The types of outcomes may be cast, for example, as number of pounds of contaminant removed, volume of water delivered, and so forth. Thus, a typical C-E outcome may be a comparison of two treatment options where one costs $X per milligram of total dissolved solids removed, and another option has a cost of $Y per milligram removed, and if $X < $Y, then that option is more cost-effective. C-E is thus a very useful approach for considering where one gets the greatest yield per dollar spent when the alternatives under consideration provide identical outcomes and only vary in cost.

C-E analysis has limitations, however, especially when alternative options have a variety of different outcomes. For example, one water supply option may cost $100 less per unit of water volume produced than a second alternative, but the second supply alternative may provide higher quality water (thus enabling saving treatment costs) or it may provide a more reliable, drought-resistant yield (i.e., providing a more consistent supply yield, even in dry years when the water is most needed). In such a comparison of alternatives that produce dissimilar types or levels of benefits, it is important to move beyond C-E and instead apply a BCA or TBL assessment.

Life-cycle cost analysis (LCCA) is another term often used and refers to the desire to capture the full spectrum of outcomes from a lifetime perspective. It is not a distinct type of analysis per se; rather, it provides a basis for how to set the boundaries—temporal and often spatial—for a BCA or similar analysis.

For example, the initial cost of building a water treatment plant does not reflect the full life-cycle costs of that facility. At a minimum, one needs to also consider the annual operations and maintenance (O&M) expenses and periodic replacement costs that accrue in operating the facility over its expected lifetime. This provides the life-cycle costs of building and operating the facility.

LCCA can be expanded in many ways to account for other types of effects across time and space. For example, a life-cycle accounting of the carbon footprint of a project might consider the energy and greenhouse gases embedded in producing the materials used to build the facility, the chemicals used in operating it, transporting all inputs to the facility, and so forth. The nature and ultimate fate of residuals should also be factored into a comprehensive LCCA.

Multiple-criteria decision analysis (MCDA) is another form of analysis often used to evaluate alternative options. It is a useful tool that can be used in group settings, because it is driven by subjective weights and scorings that are assigned by the "participants" to the various outcomes or attributes for each option under consideration. In essence, MCDA entails defining what factors or attributes to use as the basis for evaluating alternatives, and then assigning weights and scores to the factors and options.

The evaluation factors typically include cost, operational ease, yield, energy use, and so forth. In MCDA, each factor (criteria) is assigned a subjective weight, reflecting how important the scorekeepers feel that factor is relative to the others (e.g., if "cost" is more important than "operational considerations," then the evaluating group would place higher weight on cost outcomes, perhaps a 10 on a 10-point scale, whereas other factors may get weights ranging from 1 to 9). Then, each option is given a "score" for how well it performs relative to each factor (e.g., a high cost option may get a score of "2" on a 10-point scoring scale, and a relatively low cost alternative may get a "9"). Then, each option has its overall MCDA score developed by multiplying each criterion's score by its weight and adding across the criteria.

In essence, MCDA is a subjective form of BCA, where the factors important to the decision makers are considered and weighed and a final score is given to each alternative. It can produce fairly easy-to-understand results and does not entail a lot of sophisticated analysis. The main limitation is that the weights and scores are subjective and assigned by a limited group of individuals.

DEFINING IMPORTANT TYPES OF BENEFITS AND COSTS

The types and levels of benefits and costs associated with a specific water supply option are highly utility, location, and option specific. However, a review of the literature and case studies identify several types of benefits and costs that are often important when considering an alternative.

Key Benefits and Costs to Include in the Assessment

For some, a major concern regarding the development of water projects is the potential for adverse environmental or public health effects, or both. These issues and concerns certainly need to be given due consideration. Also important are the benefits generated by a water resource project, including the water supply yields, their reliability under different weather and climatic conditions, and the quality of the water.

The environmental impacts may include the following, some of which may be described as potential benefits, but some may also arise as potential costs for some projects:

- Increased ability to meet critical instream flow conditions for fish and other aquatic species and ecosystem services of concern by reducing demands on existing freshwater surface and/or groundwater supplies
- Reduced energy consumption and air pollution—such as where imported waters would be the alternative to local resources—by reducing the need for pumping large volumes of source water across great distances and gradients
- Increased protection of groundwater systems—from subsidence, reduced storage capacity, and saltwater intrusion—by reducing pumping demands on aquifers
- Increased reliability and drought relief (i.e., reducing the variability and uncertainty about the volume of water available to the community in the event of droughts or other source water–impacting events)
- Increased local control (e.g., reuse water can be viewed as a local resource, in contrast to waters imported to a community from regions and/or by agencies beyond the jurisdiction and control of the local community)

- Sustaining agricultural communities, by reducing municipal demands on waters currently applied to irrigation (e.g., by providing reclaimed water to farmers in lieu of potable supplies)
- Any cost savings associated with using reclaimed water (for example) relative to other water supply and wastewater management options (e.g., costs avoided because new waters and/or related water supply or treatment infrastructure expansion will not need to be incurred by the community or because wastewater treatment plant upgrades can be postponed while reducing wastewater discharges)

Other potential social, economic, and institutional benefits and costs to be considered (and most of which could be considered as specific subcategories under the items listed above) include

- local and regional economic effects under alternative water supply constraints (i.e., changes in local economic income and tax revenues, and employment levels, with and without alternative water supplies);
- the risk-reducing value associated with developing alternative mixes of water supply options within a portfolio approach to water supply management (i.e., diversifying the types and sources of variability in water yields and/or costs across the supply options in a community's water portfolio);
- the advantages and disadvantages of changing the level of interdependencies between the water supply and power sectors; and
- comparative risks associated with different levels of dependence on technology for water supply.

The sections that follow describe empirical evidence and a framework to enable water agencies with the means to identify and (in many instances) estimate types and likely level of all relevant benefits and costs associated with water supply options (including desalination or reuse) using a regional and community-wide perspective to better capture the full range of potential societal returns and costs from investments in new supplies.

KEY STEPS OF ALTERNATIVES ANALYSIS

A series of steps are involved in conducting and communicating a sound analysis of alternatives and include*

- defining a suitable baseline that includes recognizing how future changes will affect a utility that adheres to its "status quo" approach;
- including effects that are important, even if they cannot be readily quantified or monetized, so that these important benefits and costs are suitably identified and described qualitatively;
- recognizing and acknowledging uncertainties and omissions, and considering how these may affect the net benefit results using tools such as sensitivity analyses;
- including stakeholder involvement at all stages of the process; and
- promoting transparency and communication throughout the process to enable replication and foster trust in the outcomes.

* Much of the approach is drawn from economic analyses guidelines by Raucher et al. (2006, 2007).

A Comparative Context, With Careful Attention to Defining the Baseline

One important key to conducting a proper economic evaluation is to place water projects in a comparative context—evaluating these options against a default scenario of continuing current practices (if viable) as well as in comparison to a range of applicable water management alternatives. The key is to set up the economic analysis in a "*with* versus *without*" context in which an option can be compared to a baseline where the current practices are continued into a changing future and/or in a context where one or more options can be compared to other feasible alternatives.

Two key related points about the baseline are as follows:

1. *The future will not look like today.* The baseline needs to reflect the future, not today or yesterday. A good economic analysis is suitably oriented to a "with and without" perspective; this is not the same as "before and after." Maintaining current practices into the future will have different outcomes than those seen with current practices today (e.g., because of changing demand due to changing populations and climate, increasing costs, present options becoming foreclosed).
2. *Defining the baseline will be a challenge.* Defining the baseline may be the most contentious and difficult part of doing a good economic analysis. Expect stakeholders to hold different views about what the future will (or should) look like for a community; this will affect the assumptions that drive the analysis.

With regard to the second point, a challenge to defining the baseline is that the "with and without" context can become a place where hidden agendas or disagreements over core assumptions often arise between stakeholders and the utility or other interested parties. For instance, setting the baseline may set off a debate between the utility and stakeholders over future water demand projections (e.g., where some members of the community hold alternative views about the size and pace of future population growth, the ability to conserve enough water to meet future demands with current supplies, or about the long-term acceptability of water reuse). Therefore, it is important to carefully define the baseline, be transparent about underlying assumptions, and engage relevant stakeholders at this critical stage of the economic analysis.

Another key aspect of the baseline in a comparative economic framework is that it is important to establish and then carefully maintain a suitable accounting stance. For example, if a water reuse project enables a community to forgo (or postpone) capital and/or operating expenditures in favor of alternatives that increase the potable water supply, then these cost savings need to be included in the economic analysis. However, the side of the benefit–cost ledger that this entry gets placed on depends on the baseline scenario and what other supply alternatives are included in the assessment. Ultimately, the key is to ensure that cost savings are not double counted as a benefit for one option if it is also shown as a cost for an alternative option (i.e., the cost of option A should not also be shown as a cost savings benefit for its alternative, option B; it needs to be shown as either the cost of A or a cost-avoided benefit of B, but not both).

Steps for Conducting a Sound Alternatives Analysis

There is a series of steps that should guide an economic analysis of water resources (or other water- and wastewater-related) projects. These steps provide a logical and consistent process for the analyst to follow throughout the assessment.

The following steps make up the economic framework for analysis of alternatives and are summarized in Figure 9-1.

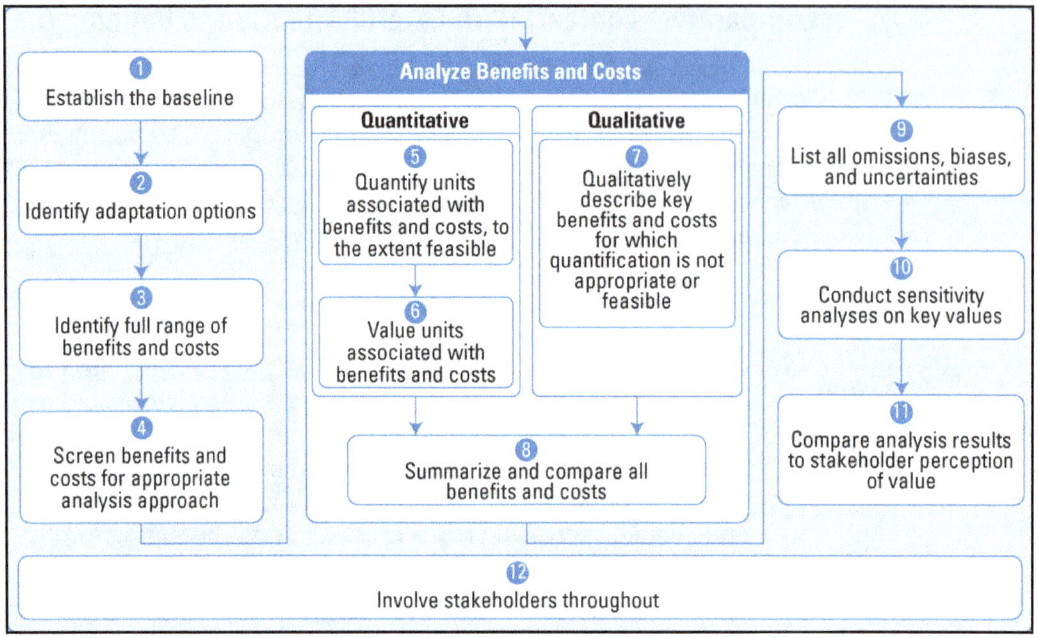

Figure 9-1 Steps in an economic analysis framework

Step 1. Establish the baseline. Define the outcomes associated with the "no action" status quo (i.e., what would happen *without* the option(s) being considered). This base case may entail doing nothing different (i.e., not pursuing a different water supply project) or undertaking already planned actions. The baseline is the mark against which changes resulting from the water alternative(s) are measured. It is important to define the scale and timing of the effects of the baseline, articulate what problems the proposed project (or range of project alternatives) is intended to resolve, be explicit about assumptions, and engage stakeholders about their perspective of what happens under a "no action, status quo" baseline.

Step 2. Identify water agency options. Identify and develop all the relevant utility options that will be compared to the baseline and to each other. It is useful to scale project options to a common size or objective (e.g., to meet projected minimum water delivery requirements). For options available at different scales, it is helpful to consider staging or combinations of options.

Step 3. Identify the full range of relevant benefits and costs for the selected option. Develop a thorough inventory of all likely costs and benefits associated with each of the project alternatives (options). Include costs and benefits beyond those faced by the utility alone or customers alone. In other words, try to identify all the benefits and costs, regardless of who is affected or where they might be realized.

Step 4. Screen benefits and costs for appropriate analysis approach. In the screening step, the analyst determines which costs and benefits can and should be analyzed quantitatively, which should be described only qualitatively, and which are insignificantly small and can be eliminated from further analysis.

Step 5. Quantify units associated with benefits and costs, to the extent feasible. In the first step of valuing a benefit or cost, the amount (quantity) of the outcome should be established. These quantity outcomes may be a volume of water delivered, amount of energy produced or used by an option (e.g., kilowatt hours), or whatever units the outcomes are most readily and meaningfully measured. It is important to match the quantity units of measurement to whatever metric is available for the corresponding dollar values (e.g., if the valuation in Step 6 uses a dollars-per-kilogallon measure, then the quantification in

Step 5 should be aimed at estimating the number of kilogallons generated or saved by an option). Ranges of quantity estimates (rather than a single-point estimate) may be used to better represent variability or uncertainty associated with resource use estimates.

Step 6. Value units associated with benefits and costs in monetary terms. Once the quantity of resource use has been estimated, a per-unit dollar value can often be assigned to the benefit or cost to reach a total value (quantity times per unit value). The per-unit values can be expressed as dollars per unit of product (e.g., dollars per volume of water produced) or dollars per unit of resource use or value added (e.g., dollars per change in cropland yields, dollars per bushel). Ranges of values may be used to better represent per-unit resource valuations. Annual benefit or cost values should be projected over the project life (either annualized or as a net present value [NPV], as per Step 8).

Step 7. Qualitatively describe key benefits and costs for which quantification is not appropriate or feasible. It may not be feasible or desirable to express some types of benefits or costs in quantitative or monetary terms (as per screening in Step 4). However, it is always important to describe these nonquantified benefits and costs in a meaningful, qualitative manner. One way to do this, in part, is by using a simple scale indicating the likely effect on net project benefits. Effects can be qualitatively ranked on a 5-point scale, ranging from –2 to +2, to reflect unquantified relative outcomes that span from very negative to very positive (e.g., a "–1" may signify an outcome with moderate unquantified costs, and a "+2" may represent a high unquantified benefit). Qualitative ratings should be accompanied by descriptions of the effect and should be explicitly carried through the analysis.

In addition, if the analyst does not wish to pursue quantitative estimates of benefits, he or she might opt to portray the analysis at this stage as a TBL analysis. This would, in essence, skip Steps 5 and 6 above, and instead focus on a systematic effort of portraying the benefits and costs in a qualitative framework.

Step 8. Summarize all present value (or annualized) costs and benefits, and compare benefits to costs. Quantitative benefit or cost projections over time (from Step 6) should be discounted to present value at an appropriate discount rate (the discount rate, which is similar to an interest rate, is described later in this chapter). The NPV of monetized benefits and costs should be summarized in one location (i.e., a summary table), along with the listing and ranking of those benefits described only qualitatively (from Step 7). It is important that one summary table includes both the monetized benefits and costs, as well as a listing and some qualitative assessment of the nonquantified benefits and costs, so that reviewers do not overlook potentially important outcomes when reviewing the empirical results. Distributional aspects should also be presented to indicate who bears the costs, as compared to who might receive benefits from the project.

Step 9. List and assess all omissions, biases, and uncertainties. All omissions, biases, and uncertainties associated with the estimated benefits and costs should be explicitly documented. The effect that these may have on the final outcome of the analysis (e.g., in terms of their likelihood of increasing or decreasing net benefits, or an uncertain direction of change in net benefits) should be noted.

Step 10. Conduct sensitivity analyses on key values. Sensitivity analyses should be conducted on key variables or benefit and cost estimates to explore and communicate the effect of assumptions, uncertainty, or natural variability. Use sensitivity analyses to identify which assumptions or uncertainties have the largest effect on the outcome of the analysis (e.g., identify which assumptions might change the net benefits of an option from positive to negative, or alter the ranking of options in terms of their relative net benefits).

Step 11. Compare analysis results with values from the stakeholder perspective. The quantitative and qualitative values resulting from the analysis and from the various sensitivity analyses should be compared with stakeholder expectation of values. This comparison of expected values to the values derived in the analysis can be informative both as a check on the reasonableness of the analysis results and as a process for working with stakeholders to realize (or at least better articulate) the values that the water resources project provides to stakeholders. This understanding of values may become the basis for cost-sharing agreements with stakeholders to share costs for a project according to the relative shares of benefits derived from the project.

Step 12. Engage with stakeholders throughout. The horizontal box on the bottom of Figure 9-1 emphasizes that stakeholder involvement should be sought throughout the project's identification and valuation process, with stronger involvement recommended at certain portions of the process (e.g., especially at the outset, and again to review and discuss findings).

METHODS FOR ESTIMATING BENEFITS AND COSTS

Numerous approaches can be used for developing estimates of the monetary value of many of the benefits and costs associated with a water resources project.

Market Price

When there is a well-functioning market for a good or service that is affected by a water resources project, the observed market price can be used as the dollar value to insert in the benefit–cost framework. Market prices are typically used for the direct costs of the project (or its alternatives), such as the cost of capital equipment, labor, and so forth. These market prices are sufficient to cover the needs of a financial analysis. However, for an economic assessment of benefits and costs, many of the important outcomes pertain to "nonmarket" goods and services. This means that there are no market prices to observe for many key outcomes (e.g., for a day enjoying an outing in a wetland or fishing on flow-enhanced streams). Thus, nonmarket valuation approaches are required for many benefits and costs.

Nonmarket Valuation

Because many of the important benefits provided by water resource management involve nonmarket goods and services, monetary values need to be derived using various well-established methods developed by economists for "nonmarket valuation." These nonmarket valuation approaches can help develop dollar estimates for some important types of water resource–related benefits and thereby help decision makers and the public better recognize the value of a water resource option. These nonmarket valuation methods are summarized below.

Primary methods. Many goods and services associated with water-related effects are not traded in markets. For example, rarely are there well-defined markets for many water-based recreational activities. There are two main approaches that economic researchers can use to estimate nonmarket values via primary research. These are known as *stated preference methods* and *revealed preference* methods. Stated preference methods are survey based and include contingent valuation and conjoint analysis, and revealed preference methods include travel cost and hedonic modeling (see summaries in Table 9-1).

Revealed preference methods are based on observing individuals' behavior and associated voluntary cost bearing to infer the value of a nonmarket good or service. While there may not be active markets to buy and sell days of outdoor recreation, there are often costs that individuals incur to undertake direct use activities. For these types of uses, incurred costs can often be applied to develop proxy "prices" for the activity, and that information is used in developing the demand curve, and thus value, of recreation-related services. This approach uses observations on people's behavior, or their associated expenditures, as indications of "revealed preferences" for the good. Methods have been

Table 9-1 Primary economic valuation methods for nonmarket goods and services, comparative advantages and disadvantages

Method	
Travel Cost	
	Pro: Uses observed tourist and recreation trip-taking behavior
	Con: Measures use values only; often expensive and time-intensive to collect adequate data
Hedonic Pricing	
	Pro: Uses observed housing, property, or labor market behavior to infer values for environmental quality changes
	Con: Measures use values only, requires extensive market data, and assumes that market prices capture the environmental good's value
Contingent Valuation	
	Pro: Only method that can estimate nonuse values; also can estimate use values
	Con: Time-intensive and expensive to implement, challenges in framing survey questions to elicit valid responses, and potential response biases
Conjoint/Stated Choice	
	Pro: Similar to contingent valuation, except respondents are surveyed about a set of choices instead of a single willingness-to-pay question
	Con: Time-intensive and expensive to implement, challenges in framing survey questions to elicit valid responses, and potential response biases

developed, and are discussed below, that use these revealed preferences to develop estimates of the value of nonmarketed goods and services such as many water uses.

For other activities, where there is no direct use of the resource and thus no behaviors or expenditures available as a measure of people's preferences, methods have been developed to directly elicit preferences and estimate value. These direct methods are often described as "stated preference methods" because they most commonly elicit value through direct statements on value rather than using behavioral observations or expenditures to infer value.

The most commonly revealed preference methods are the *travel cost method* and the *hedonic pricing method*. The travel cost method is used to value recreational uses of natural and environmental resources. The hedonic pricing method can be used to value a wide variety of factors that influence observed prices and is often used to infer the value of environmental goods.

Two commonly stated preference methods are the *contingent valuation method* and the *conjoint/stated choice method*. The contingent valuation method can value not only direct use values but also nonuse (e.g., existence and bequest) values for natural and environmental resources. The conjoint/stated choice method asks for a ranking of choices instead of an answer to one willingness-to-pay question and can also be applied to derive estimates of either use or nonuse values.

Secondary methods. Primary research is often expensive to execute correctly and often is not feasible because of budgeting, scheduling, and other constraints. It is often more practical to turn to secondary methods (described below) and to use an approach that helps identify the critical values in the BCA. If a particular value is identified as critical, then perhaps it will become desirable to invest in a primary research study to more definitively determine that value.

Benefits transfer (BT). This is an expeditious method for valuing nonmarket environmental resource services. The BT approach involves taking the results of existing valuation studies and transferring them to another context, for example, a different geographic area or policy context (Bergstrom and De Civita 1999). Under suitable circumstances, estimates for use or nonuse values may be derived, for example, using BT by applying an annual willingness-to-pay estimate per household to all the households in the geographic area in question with the same use or nonuse motives for the resource.

There are numerous challenges and cautions to consider when using BT. While it is relatively simple to develop a BT-based-value monetary estimate of many types of benefits (e.g., there is a large amount of literature on user-day values for recreational experiences associated with improved surface water or wetland conditions), there are numerous ways in which the approach can generate potentially inaccurate (and misleading) results, even when a well-intentioned and objective analysis is being attempted. The most significant challenges to the accuracy and credibility of BT-generated findings are that there often are important differences between what type of natural resource conditions were studied in the primary empirical research (i.e., the study context for the published monetary estimate) and the water resource context and site to which an analyst may wish to transfer the results.

BT is commonly used in economics, and literature on correctly applying this method is well developed (e.g., Rosenberger and Loomis 2003). Federal agencies—US Environmental Protection Agency (USEPA) and the US Office of Management and Budget (USOMB)—issued guidelines for economic analysis that discuss how and when BT should be applied (USEPA 2000, USOMB 2003). When implemented correctly, the BT approach is accepted as a suitable nonmarket estimation method for estimating the use and nonuse benefits of changes in the level or quality of environmental resources, especially when used cautiously and transparently, and with a recognition that the estimates are not intended to be precise. However, primary research is broadly considered a far better alternative when time and resources allow.

Societal revealed preference. Nonuse values may be deduced (under limited circumstances) using voluntarily incurred restoration-based costs as a proxy for the value of the change in resource conditions. For example, for threatened or endangered species, the costs of voluntary or consensus-driven restoration programs and the costs imposed by various widely endorsed resource-use restrictions may indicate the revealed preference value of restoring species populations to sustainable levels.

Avoided costs (cost offsets). Avoided costs may be an important part of valuing the range of benefits likely to be generated by a water resource project. For example, developing a local reclaimed water program to supplement the potable supply may help avoid or defer the expense of relying on expensive imported water and/or postpone the need to expand the potable water treatment facility. Using NPV analysis will allow for a fair comparison of benefits accrued in different years to be made.

There are potential issues, however, to be alert to when using avoided costs as a proxy for benefits values. Avoided costs can be used as measures of benefits when they would actually be incurred in the absence of the water resource project. There is a potential for double-counting avoided costs in a BCA, and analysts need to be alert to this possibility when defining the baseline (and determining that costs of some options do not simultaneously appear as cost savings benefits of their alternatives).

Replacement costs. In some cases, a lower bound value for a lost resource can be estimated according to the costs necessary to replace the resource. For example, with loss of wetlands, an estimate can be derived for the cost to replace that habitat. However, using cost-based measures as proxies or lower bounds for values can be tricky. Costs should only be used in this manner if they have been incurred voluntarily or through a consensus-based process. Otherwise, it is inappropriate to assume that costs also reflect values.

Response cost: averting or mitigating behavior. The averting-behavior approach examines the expenditures people make to avoid damages resulting from environmental degradation. The mitigation approach examines the expenditures people make to correct a problem after the potential effect has occurred. This could include measures such as installing water purification or filtering devices in the home to improve tap water taste because a low aesthetic-quality source water is being used by the utility.

DISCOUNTING AND PRESENT VALUES

Benefits and costs from water supply planning and management projects often occur as a stream of values over time. In addition, that stream may change in magnitude over time. Most options usually have large capital costs that are paid for either up front or, more likely, over an amortization period at the beginning of a project. Benefits, however, may accrue over the economic life of the project, which for some projects can be substantially longer than the amortization period. Values that occur in different time periods need to be adjusted to their comparable "present values" to compare them or make calculations with them.

There are two interrelated factors to consider when comparing values from different times—inflation and the "time value of money." When inflation is accounted for in recording or projecting values over time, then the values are said to be in "nominal" terms. Many financial analyses are conducted in nominal dollars. However, for economic analyses, benefits and costs are normally *not* entered in inflation-adjusted dollars. Instead, the use of "real" (i.e., not inflation-adjusted) dollars makes analyses easier and keeps inflation-related projections from clouding the analysis. In real dollars, a dollar today has the same purchasing power as a dollar 10 years from now.

The second factor to account for in comparing values over time is because most people prefer a dollar today more than an inflation-adjusted dollar available in the future. Most prefer to have a real dollar today instead of a real dollar in the future because they prefer to use that dollar to consume today or they prefer to invest that dollar today to yield a future return. This preference for near-term consumption over deferred consumption is referred to as the "social rate of time preference" or the "time value of money." This social rate of time preference is the real (i.e., inflation free), net of tax, and risk-free rate of interest that would need to be paid to a person to entice them to consider delayed receipt of a real dollar.

The annual rate at which present values are preferred to deferred values is known as the "discount rate" (and is similar to an interest rate). The greater the preference for immediate benefits (time preference) or the greater expected rate of return on other investments today (known as the "opportunity cost of capital"), then the greater the discount rate. The discount rate can be expressed in nominal or real terms. A real discount rate is the nominal discount rate with the inflation rate subtracted out. The key is to use a real discount rate when analyzing dollars in real terms and a nominal discount rate when analyzing values in nominal terms.

Economic theory suggests that in a world with no taxes, no financial transaction costs, and zero risk, there would be a clear signal about what discount rate to use. If consumption today would come at the expense of investments in the future then the opportunity cost of capital should be used to discount the stream of future benefits and costs. In that case, the discount rate should be equal to the rate of return that could be earned by investing the money. If inflation is expected to be 4 percent in the future and there is a 3 percent risk-free real return on capital, then the real discount rate would be 3 percent, and the nominal discount rate should be 7 percent (3 percent + 4 percent). If instead the use of funds or resources today predominantly displaces future consumption (instead of investments), then a social rate of time preference is more suitable as the discount rate.

There are philosophical and practical aspects to the choice of discount rate, and there is not always general agreement among economists or policymakers about the correct discount rate to apply to evaluating projects. For BCAs of water supply and other water management projects, which are generally investments made for broad public benefit, it may be most appropriate to use a real, net of tax, social rate of time preference as a real discount rate to convert all values to their present worth. However, justifications can be made for a range of rates, from a zero discount rate to a discount rate reflecting the private cost of capital.

- The argument for a zero discount rate is that discounting distorts project benefits that may occur far into the future, and thus affect future generations, or that include irreversible outcomes (e.g., species extinctions).
- Others suggest that the discount rate should reflect prevailing interest rates on low-risk bonds because such risk-free, net-of-tax rates best reflect the rate of social time preference. This might be reflected by the real cost of capital to municipal agencies in raising capital through bonds or the cost of long-term federal government bonds.
- Another argument is that for projects that will be paid for through rates, the cost of capital to ratepayers is the appropriate measure. Depending on whether the utility is a public or a private entity, the cost could range from 3 to 12 percent in nominal terms.
- The argument for using the private cost of capital is that the project's funds might be otherwise invested in private ventures, and therefore reflects the true opportunity cost.

Various governmental entities have specified discount rates to be used in analyses. The USOMB regularly updates discount rates in Appendix C to its Circular Number A-94 on *Guidelines and Discount Rates for Cost Benefit Analyses of Federal Programs* (USOMB 2014). USOMB recommends using real interest rates on Treasury notes and bonds matched to the project time period for the real discount rate. The real interest rate on a 30-year note as of February 2014 was 3.5 percent. Federal water resource agencies are also directed to use specific rates to evaluate water project alternatives by the Federal Code of Regulations, Plan Formulation and Procedures; and for federal fiscal year 2014, the general planning rate is also 3.5 percent (USDA 2014).

USOMB also mandates that federal agencies apply a 7 percent real rate of discount when evaluating the costs and benefits of federal regulatory actions (USEPA 2000), although other rates are often used in sensitivity analyses (and a 3 percent real rate is typically used by the USEPA to reflect the social rate of time preference).

To compare streams of value over time from different projects, the stream of values for each project is discounted to "present value" using the discount rate. If both benefits and costs are involved, the present value of the costs is subtracted from the present value of the benefits to get the NPV of the project. If the NPV of a project is greater than zero, then the present value of the benefits is greater than the present value of the costs. The NPV of different projects can be compared if they are adjusted to be in the same year's dollars. Comparison of NPV of projects allows fair comparisons of project values regardless of possible differences in the timing of benefits and costs for each project.

DEALING WITH UNCERTAINTIES AND DATA GAPS

No process contains all information, so managers need to deal with the inevitable array of uncertainties and data gaps that arise when trying to estimate relevant benefits and costs. This includes the need for, and value of, techniques such as sensitivity analysis and scenario analysis, break-even analysis, and other strategies that can provide informative and

Table 9-2 Sensitivity analysis applied to discount rate (thousands of dollars)

Discount Rate (%)	Monetized Benefit	Cost	Monetized Net Benefit (NPV)
0	49,000–51,500	30,000	19,000–21,500
3	39,500–41,700	26,000	13,500–15,700
6	29,500–34,000	22,000	7,500–12,000
9	15,950–21,300	16,000	(50)–5,300
12	8,500–14,000	11,000	(3,500)–3,000

transparent results to stakeholders and decision makers, even in the face of considerable uncertainty.

What is critically important for the benefit–cost framework is that it provides a systematic and objective avenue for accommodating potential or estimated risks, including the often-considerable scientific uncertainty surrounding those risks. It is often important to qualitatively discuss effects that may be difficult (or impossible) to quantify in a reliable or meaningful manner, and the targeted use of sensitivity analyses can help reveal whether these types of effects might alter how a water supply option ranks relative to its alternatives. The BCA approach also provides a comparative framework, so that a broad range of feasible alternatives—spanning a variety of options—can be assessed relative to each other, and their types and relative levels of risk can be suitably portrayed.

What Is Sensitivity Analysis and How Does It Help?

In many cases it will be useful to explore the effect of uncertainties or key assumptions (such as the choice of discount rates) through the use of sensitivity analysis. Sensitivity analysis involves systematically changing the value of some key input variables to see how it affects the outcome of the analysis. The change in results with the change in inputs can show how important the effect of uncertainty in a particular variable is to the outcome. Sensitivity analysis is often performed by varying a particular input by equal amounts—both greater to and less than the current value.

For example, if a discount rate of 6 percent has been chosen for the main analysis, that value might be varied in increments of three percentage points from 0 to 12 percent for the sensitivity analysis. Table 9-2 shows an example of a sensitivity analysis for the discount rate applied in this fashion to an example project.

How Should Uncertainty Surrounding This Analysis Be Handled?

In an ideal situation, data would be available to statistically estimate confidence intervals for benefit or cost estimates. When this is possible, confidence intervals for estimates should be noted. However, statistically estimated confidence intervals are most often not possible. When it is possible with available data, ranges should be developed for an estimate by stating the upper and lower bounds. When bounding an estimate is not possible, one can at least characterize uncertainty qualitatively by describing the sources of uncertainty and stating whether an estimate developed is likely to over- or underestimate the true value (see omissions, biases, and uncertainty, which is Step 9 of the framework process, described earlier in this chapter).

Imprecision in Estimates

It should be noted that there are two main sources of imprecision in estimates of values. One is *variability*—the natural variations in an estimate due to its properties or the forces acting on it (e.g., rainfall can vary by season and from year to year). The other is *uncertainty* about an estimate due to a lack of knowledge on the part of the analyst about the true value (e.g., is reliability the value of an improved water supply). Both variability and uncertainty can lead to imprecision of estimates, and are both reasons why estimates should typically be represented with a range of values instead of just a single value. A single "best estimate" or "mean value" can be used, but the range of possible values should be identified and explored using sensitivity analysis. Using a range of values instead of only a single estimate can avoid giving any impression that the analysis is tilted toward a desired outcome.

Risk and Uncertainty

A related concept to uncertainty is *risk*, which, for the purposes of this discussion, can be thought of as reflecting the probability and consequences of an event. For example, prolonged and intense rainstorm events carry risks in terms of the quality and quantity of surface source water. Drought and storm events reflect variability in precipitation from year to year (i.e., there is variation in the level and pattern of rainfall over the years), and there are uncertainties about when the next drought or extreme wet period will occur or how severe it will be. The risk associated with a highly intense wet period is that there is some probability that in a given future year(s), there will be elevated turbidity in the source waters, and the consequence is that in those years (or weeks) there may be a strain on utility surface water treatment capabilities to comply with drinking water standards. Not all water resource options bear the same types and/or levels of risk, and these differences are important to articulate in the analysis. In some instances, monetary values may be assignable to differences in risks across options (but in many cases, monetary measures of differences in risk will not be readily monetizable using available literature).

Scenario and Sensitivity Analyses

Scenario analysis and sensitivity analysis are two important tools to help understand the effect of uncertainty. By examining different scenarios with different values from the range of uncertainty for key variables, the analyst can examine whether the uncertainty in the underlying variables is important to the ultimate outcome of the analysis. Knowledge of whether uncertainty regarding key variables is likely to affect the outcome of the analysis or the decisions to be made can help focus future research efforts on the most productive topics. Probability-driven approaches are also valuable, and where a distribution of possible outcomes can be formulated, one useful approach is to conduct a Monte Carlo simulation.

CONCLUSIONS

In evaluating water resource projects, developing a clear and meaningful comparison of the alternatives is important. One alternative is always the "do nothing" status quo baseline, reflecting a future without any new water resource option being implemented. There are various approaches that may be usefully deployed in an analysis of alternatives, and some of the key points to consider regardless of the method applied include the following:

1. Compare fairly, so that decision makers can focus on true alternatives and their respective abilities to meet key objectives (and/or understand key ways in which the alternatives may yield different types or levels of key outcomes).
2. Pay careful attention to identifying and describing all the important benefits and costs (pros and cons) so that nothing of significance is omitted (even if the outcomes cannot be quantified or monetized).

3. Identify clearly where there are large uncertainties, unquantified benefits and costs, externalities, or the factors that are important to consider for decision makers.
4. Articulate key uncertainties and use sensitivity analyses and other suitable methods to assess and communicate the size and implications of those uncertainties.
5. Recognize that the analysis is a tool to help inform the decision-making process (it is not a rule to determine a decision).
6. Keep the analysis and the results as transparent and replicable as possible so that they can be effectively communicated with decision makers and stakeholders.

REFERENCES

Bergstrom, J.C., and P. De Civita. 1999. Status of benefit transfer in the United States and Canada: Review. *Canadian Journal of Agricultural Economics*, 47(1):79–87.

Elkington, J. 1998. *Cannibals With Forks: The Triple Bottom Line of 21st Century Business.* Gabriola Island, BC: New Society Publishers.

Raucher, R.S., K. Darr, J. Henderson, R. Linsky, J. Rice, S. Bahman, and C. Wagner. 2006. *An Economic Framework for Evaluating the Benefits and Costs of Water Reuse.* Final Project Report and User Guidance. 03-006-02. Alexandria, Va.: WateReuse Foundation.

Raucher, R.S., D. Garvey, K.C. Hallett, J. Henderson, and C. Wagner. 2007. *An Economic Framework for Evaluating the Benefits and Costs of Biosolids Management Options.* Report 04-CTS-2. Alexandria, Va.: Water Environment Research Foundation; and London, UK: International Water Association.

Rosenberger, R.S., and J.B. Loomis. 2003. Benefit transfer. In *A Primary Non Market Valuation.* Edited by P. Champ, K. Boyle, and T. Brown. Boston: Kluwer Academic Press.

USDA (US Department of Agriculture). 2014. *Rate for Federal Water Projects.* Natural Resources Conservation Service. www.nrcs.usda.gov/wps/portal/nrcs/detail/national/technical/econ/costs/?cid=nrcs143_009685 (accessed May 20, 2014).

USEPA (US Environmental Protection Agency). 2000. *Guidelines for Preparing Economic Analyses.* Washington, D.C.: USEPA.

USOMB (US Office of Management and Budget). 2003. *Regulatory Analysis.* Circular A-4. US Office of Management and Budget. https://www.whitehouse.gov/sites/default/files/omb/assets/regulatory_matters_pdf/a-4.pdf (accessed June 2, 2016).

USOMB (US Office of Management and Budget). 2014. 2014 Discount Rates for OMB Circular No. A-94. Memorandum from Director S. Burwell to Heads of Federal Departments and Agencies dated February 7. www.whitehouse.gov/sites/default/files/omb/memoranda/2014/m-14-05.pdf (acessed May 20, 2014).

Chapter 10

Integrated Water Management Plan

The issues discussed in this manual should allow the planner to develop and implement a comprehensive work plan that responds to technical and institutional questions that must be addressed before deciding how to develop new water supplies. However, planners are expected to use the parts of the manual that apply to their situations. Key elements of a water resources plan should include the following, most of which are addressed in the preceding chapters:

- Plan the plan; that is, have the plan thought out with clear goals in mind.
- Engage key stakeholders.
- Know state and federal laws and regulations.
- Identify alternatives.
- Estimate future water demands.
- Evaluate water quality.
- Conduct a source water assessment and corresponding protection plan.
- Identify evaluation criteria.
- Conduct a screening-level evaluation of alternatives.
- Select appropriate methods for detailed evaluations.
- Evaluate alternatives and document the evaluation process.
- Develop recommendations.
- Prepare for monitoring, evaluating, and updating of the plan.

Proper water resources planning includes developing a sustainable program for water supply for the utility. With respect to water resources, *sustainability* can be generally defined as maintaining the rate of water consumption and overall water quality in perpetuity. Sustainability is more difficult to achieve than sometimes anticipated because of the importance of water to locales and regions, changing population and productivity of water sources, increasing costs and regulations, and numerous other factors.

The primary purpose of a domestic water system is to provide safe and reliable water service to its customers. Unfortunately, utilities and regulatory agencies are often responsible for defining how to provide safe and reliable water service in the context of the larger questions that may drive local decisions and alternative analyses.

DEVELOPING A PLAN

Planning is a public forum activity that provides the public with an indication of what may be expected in the future. Planning is often best accomplished by creating a vision for the future—where the community wants to be. A vision allows elected officials, the public, ratepayers, and utility staff the opportunity to outline the need for programs or to make revisions to the manner in which business is conducted. The preferred method of developing that future direction is through the planning process. As a result, planning should be undertaken on a regular basis in an effort to anticipate needs, clarify organizational goals, provide direction for the organization, and to communicate each of these aspects to the public. This process is particularly important when planning for water resources needs because identifying and implementing future water resources opportunities takes time. With water and wastewater utility systems, it is imperative to have ongoing planning activities, because many necessary improvements and programs take many years to implement. Long-term water resources planning horizons are commonly at least 20 years, and many are looking 50 years out because of the potential challenges in securing water and funding. Without a long-term plan to accomplish future needs, a utility could suffer errors in direction, build unnecessary or inadequate infrastructure, or pursue programs that later are found to provide the wrong information, level of service, or type of treatment.

The types of goals and objectives included in a plan may vary depending on the type of plan to be developed. Several types of plans may result from the planning process, as outlined in Bloetscher (2011):

- Strategic plans—action oriented for management-level decision making and direction
- Integrated water resources plans—actions for utility management to tie all parts of the system together
- Facilities plans—for specific facilities and to obtain State Revolving Fund loans support
- Master plans—to support capital improvement programs

The latter two are beyond the scope of this manual, but the first two are important for long-term planning for water resources.

Strategic Plans

Strategic plans are action-oriented, management-level documents that set a vision for the community and offer a series of strategies and action steps to be taken to reach the desired goal. These plans are typically in 5- or 10-year increments, extending 20–50 years in the future. Substantial changes to current operating and management practices may be required to meet the plan's objectives. Corporations often use strategic plans to help guide new products into the market or to develop strategies to expand their share of the market. Utilities are no different. Strategies and action steps should include not only utility issues but how the utility issues may affect, or be affected by, the activities of other sectors of the local economy and local governments. Interagency cooperation is a key part of the strategic plan. Details are often limited for a strategic plan, because the focus is on the macro-issues; details are left for those implementing the policy directives. A strategic plan can be brief or extensive.

Integrated Water Resources Plans

The concept of an integrated water resources plan for a utility system is likely developed in the implementation stage of a strategic plan, and it requires more detail, including the full scope of the operating environment, available resources, and ecological system (see Bloetscher 2011). Integrated water resources planning involves reviewing alternatives for the provision of water and wastewater over the long term and providing service to those areas that will be in need of such service.

One area of focus of an integrated water resources plan is developing the alternatives to meet current and future needs for water. The integrated water resources plan uses demand-side water management in an effort to find a balance between developing new sources and conserving current resources. Integrated water resources planning identifies the traditional, master planning elements, which require a focused evaluation of current and future demands, coupled with determining additional facility requirements, and when they will be needed. These future facilities generally represent major capital investments and additional operation and maintenance costs. In addition, the integrated water resources plan considers social and environmental impacts of alternatives, both beneficial and adverse, in the analysis of alternatives.

DEVELOPING THE PLAN

As noted previously, no two water resources plans are alike. Regardless of the type of plan resulting from the planning process, planning in the water resources field involves sequential components (Dzurik 2003), which are addressed in the chapters of this manual:

- Developing goals and objectives
- Collecting and analyzing data
- Identifying the problem
- Clarifying the problem and ancillary issues
- Identifying reasonable alternatives
- Analyzing alternatives
- Recommending actions
- Developing an implementation program
- Monitoring results

This manual follows a stepwise planning process. The intent in setting up the order for the chapters in this manual was to discuss policies and laws, then identify a comprehensive list of alternatives up front (chapter 4) as might be needed, for example, for an environmental impact statement with no evaluation or assessment of the alternatives at that point. For environmental permitting purposes, a comprehensive list of alternatives will be required. Through the evaluation process (chapters 5–9), this comprehensive list can be screened to determine which alternatives are reasonable. Concurrently, the definition of need would be defined (chapter 5), and collection of data other than water quantity would be conducted (chapters 6 and 7) to use in the evaluation of alternatives (chapters 8 and 9). Chapter 9 involves decision making using the assessment results.

Developing Goals and Objectives

Chapter 2 provides an overview of the process of setting goals and developing work plans to involve the public in the water resources planning process and public participation techniques (and the important role it can play in water resources planning). Initially, there must be a catalyst for the plan, which can include a variety of issues such as forecasted changes in demand, changes in supply, changes in water quality, competition from other water users, policy decisions like water conservation and water quality requirements, and the operational needs of the utility. These factors will define the information needed.

Before many problems can be identified and solutions sought, a series of relevant goals and objectives should be developed. The goals and objectives set the expectations that the community will have for its utility in the long and short term. Goals are used as a means to evaluate the alternative solutions and their effect on the community. As a result, those engaged in the process will understand how the challenges and their solutions will be analyzed. Therefore, if the goals and objectives do not adequately outline the intent of the constituency, the solutions may fail to adequately address the problem needs. Although defining the goals and objectives sounds easy, this may be an area of conflict. Common areas of conflict include economic issues, conflicts between environmental and water use needs (native salmon populations and urban water use in the Northwest are an example), and conflicts between water users. Goals and objectives are general in nature but clear in application.

In addition, the goals and objectives should enable those engaged in the process to understand not only the feasibility of a given option to address long-term challenges but how reliably the option will perform. For example, finding a surface water supply to address a potential long-term deficit is great, unless the reliability of that new supply fluctuates or is highly weather dependent. Groundwater supplies may appear to be an easy solution to surface water woes, but many groundwater supplies do not recharge locally and therefore have limited productivity. The reliability of the potential "solutions" may not meet the long-term reliability for the utility or the uncertainties may be large. Such considerations should be part of the analysis.

Collecting and Analyzing Data

The data collection and problem identification steps may vary depending on the utility, its age, and challenges. Data include trend analysis on water quality, reliability of water supplies, changes in demand patterns, infrastructure reliability, history of potential conflicts, and the regulatory horizon. Data collection will help clarify the scope of the problem. The initial assumption of what the scope of the problem is will likely be modified as a result of new information. The data collected should include a projection of the future. Climate effects, such as loss of storage capacity in glaciers, changes in rainfall patterns, or sea level rise, are examples of potential changes to the future water supply and utility operating condition that should be accounted for when collecting data. Data to be gathered are outlined in chapters 4, 5, 6, 7, and 8. Forecasts should be made using the data and appropriate variables to meet future needs. Keep in mind that demand projections are as much an art as science, and need to be updated as conditions change. Therefore, long-term demand projections should be understood as having significant amounts of uncertainty attached to them. Identifying timelines for water supply shortfalls is the goal for this part of the process.

Water uses and water quality data are required. Protecting the water source is an ongoing issue that has long been used by water systems, local governments, and states to protect and ensure continued viability of their water supplies.

Identifying and Clarifying the Problem and Ancillary Issues

Although separated into two steps by Dzurik (2003), these two steps are combined here because they basically form a feedback loop. As data are gathered and synthesized, it is hoped that the scope of the needs and potential challenges will become more focused. Additional and ancillary issues may be identified that relate to potential future issues, for example, competing new users, environmental conditions not previously noted, and climate issues. To address the ancillary issues, more data may be required to further clarify the problem. Once the initial data and problems have been identified, the goals and objectives should be reviewed and refined as needed. The feedback loop could take considerable time because some of the data may need to be collected and developed, as opposed to having been collected previously and just needing synthesis.

Once the additional data have been collected and public discussion ensues, crystallizing the assessment of needs should occur. Crystallizing the problem to be solved will generate more obvious and poignant solutions, thereby saving time and dollars. This phase of the planning process will likely indicate a series of solutions that can be evaluated. Defining the need will primarily be derived from the findings of the water demands discussed in chapter 5.

Identifying Alternatives

Chapter 4 identifies a series of alternatives for developing, expanding, or maximizing a water resource portfolio. Alternatives must consider the goals and objectives defined above, economics limitations of the utility, environmental effects caused or solved by the alternatives, the impact of operations and maintenance, operational complexity, and regulatory impact. The universe of solutions should be developed, possibly through a brainstorming exercise among engaged parties. Utilities should include all potential water supply options in their alternative analysis. For example, treating wastewater with reverse osmosis membranes for groundwater recharge may appear to be an overly expensive option to consider and might be summarily rejected, yet the reliability and potential increase in water supply to a resources-constrained basin may prove far more viable than the initial assessment may have indicated. On the other hand, residential reclaimed water, while appearing to be a good option, may be prohibitively expensive because of the costs of running a parallel nonpotable water distribution system.

Some poor or fatally flawed alternatives may be quickly eliminated from the analysis. Poor alternatives generally fail to meet the planning goals and objectives. Likewise, while no bad alternatives should ever be presented in a public forum, options should not be summarily rejected. Unilateral decisions are fodder for political conflict, which is nonproductive and distracting when solving the complex problems facing utility systems. Cost should not be a factor for immediate elimination.

Water quality should be a consideration in the evaluation process. Current water treatment technology allows virtually any water quality to be treated successfully. At the same time, quality issues may play a role in evaluating reliability and local treatment expertise, so quality should be evaluated with respect to costs, reliability, and other factors, not of quality alone.

Analyzing Alternatives

The process for analyzing alternatives is presented in chapters 8 and 9. An alternatives analysis often involves comparing financial, social, and environmental aspects of the alternatives. The social and environmental aspects of an alternative may be difficult to quantify and monetize, yet they may have the effect of eliminating an alternative because of the damage that may occur in the service area. Therefore, measuring the positive and negative effects of the project or program, as compared to the current condition, can be instructive but not easily quantifiable in dollars. For example, this is evident where a community is dependent on salmon fishing and an alternatives evaluation has determined that a dam is the most cost-effective alternative for future water supply. That conclusion may not recognize that the dam may affect the salmon industry, affecting jobs and the economy, and making the additional water supply unnecessary. Clearly, this is an economic decision with unintended negative consequences that might have severe effects outside the present or annual worth analysis (Bloetscher 2011).

The options must address the identified water supply gaps and needs from the problem statement above. If an option fails to address the water supply needs, gaps, or risks as noted previously, the option can be eliminated. However, many options may appear to score poorly but may need to remain in consideration for future inclusion. Such solutions may come in handy in the future if a preferred solution may fail for unforeseen reasons. As a result, it is clear that local economic and social issues, and return or in-stream flows, are all major considerations in water resources planning in addition to costs. The effects of alternatives should be considered both locally and regionally. For example, the cost for source water protection to help maintain water quality in the watershed may be offset by the savings of treating higher quality raw water supplies.

Recommending Actions

The final set of recommended solutions should be the alternatives that best meet the goals and objectives of the utility as defined in the goal-setting process, realizing that no one solution may meet all the objectives. The choice of options should be as inclusive as possible, not limiting. An assessment of risk and reliability needs to be embedded in the planning efforts. The robustness of the effort will be measured by the way it addresses risk scenarios. The recommended solutions should be adopted by the governing body as the "plan." Adopting the plan provides staff with direction to proceed with the implementation phase. In many cases, the approval will permit staff to begin immediately developing policies and procedures, and pursuing capital project construction (Bloetscher 2011).

Developing an Implementation Program

Implementing the recommended alternative occurs after the plan has been adopted. Too often, plans are adopted but never implemented, which wastes time, money, and the efforts of staff and governing bodies, and possibly erodes the public trust. Integrated resources plans are designed to be action oriented as opposed to the policy-oriented strategic plan. Many ongoing efforts will focus on the implementation issues and modifications to the plan as a result of the challenges.

Plans should be designed as living documents, which assume that changes can be made to address challenges as they arise. Many plans include imperfect data and information; an option may prove unfeasible in a given area, eliminating that option from further consideration, or legal issues may cause a delay in an option that requires reconsideration of other options. Therefore, utilities should expect that this might occur throughout the life of the plan, and strategies must be developed that are adaptable to changes in the plan's underlying assumptions for future conditions.

Monitoring Results

Monitoring the implementation and the results of that implementation are important to ensure that the chosen course of action meets the goals and objectives for which the plan was developed and that the problems identified initially are solved. The monitoring phase is where modifications to the course of action or the programs and policies can be implemented to maximize the likelihood of success. The monitoring phase also permits the utility's staff to interact with the governing board and the public in a positive manner.

CASE STUDY: CITY OF SDPUD 2012 LONG-RANGE WATER RESOURCES PLAN

San Diego has a reputation of being one of the country's most desirable places to live and conduct business because of its climate, economy, and high quality of life. It is the eighth largest city in the United States and the second largest in California. San Diego, located on the coast of the Pacific Ocean and adjacent to the Mexican state of Baja California, is an international city—economically and culturally. The city is known worldwide as a prime tourist destination and is home to important industries such as telecommunications, biotechnology, software, and electronics. It boasts one of the largest defense and military complexes in the world, and its commercial port and location along the United States–Mexico border results in significant international trade and commerce.

Although San Diego is located in a semi-arid coastal climate with an average of 10 in. (25 cm) of rainfall annually, it has successfully provided a reliable water supply to its residents for more than 100 years. The city covers approximately 340 mi^2 (880 km^2) and stretches nearly 40 mi (64 km) from north to south. It is a city defined by varied landscapes—oceans, mountains, mesas, canyons, and estuaries. This varied topography requires sophisticated and innovative water and wastewater systems.

The City of San Diego Public Utilities Department (SDPUD) manages one of the largest water storage, treatment, and delivery systems in the United States. The water system extends more than 404 mi^2 (1,046 km^2) with water deliveries on the order of 200 million gallons per day (mgd), or 757 million L/day. SDPUD also has a separate recycled water distribution system that currently extends more than 80 miles (129 km) and serves an annual average of approximately 7 mgd (26.5 million L/day) of water demands within the city and 4 mgd to three wholesale customers. In addition to securing water supplies for more than 1.3 million residents, SDPUD also manages a regional wastewater system that serves 2.2 million residents in San Diego County with a mission to "… ensure the quality, reliability, and sustainability of water, wastewater, and recycled water services for the benefit of the ratepayers and citizens served." The challenge for the City of San Diego (City) is to excel in its mission in the long term, given the uncertainty of future economic conditions, climate change, regulatory and legal requirements, and reliability and cost of current water supply sources.

Service Area Background

The City's combined retail- and wholesale-serviced population is expected to increase from about 1.41 million in 2015 to more than 1.69 million in 2035, which is about a 20 percent increase in 20 years. SDPUD forecasts water demands using an econometric approach that is based on several drivers, including socioeconomics, weather, household density, employment, and conservation. The projected growth in water demands from 2015 to 2035 is 17 percent, which is lower than the expected population growth during the same period because of water-use efficiency.

The City purchases approximately 85–90 percent of its water from the San Diego County Water Authority (SDCWA), which is approximately 72.4 percent of the overall resource mix including conservation and reuse. The SDCWA is a wholesale water agency that provided approximately 417,000 AFY, or acre-ft per year (514.4 MCM/year, or million cubic meters per year), of imported water to its 24 member agencies in San Diego County in fiscal year 2011. The SDCWA, in turn, gets most of its imported water from the Metropolitan Water District of Southern California (MWD), which delivers water to

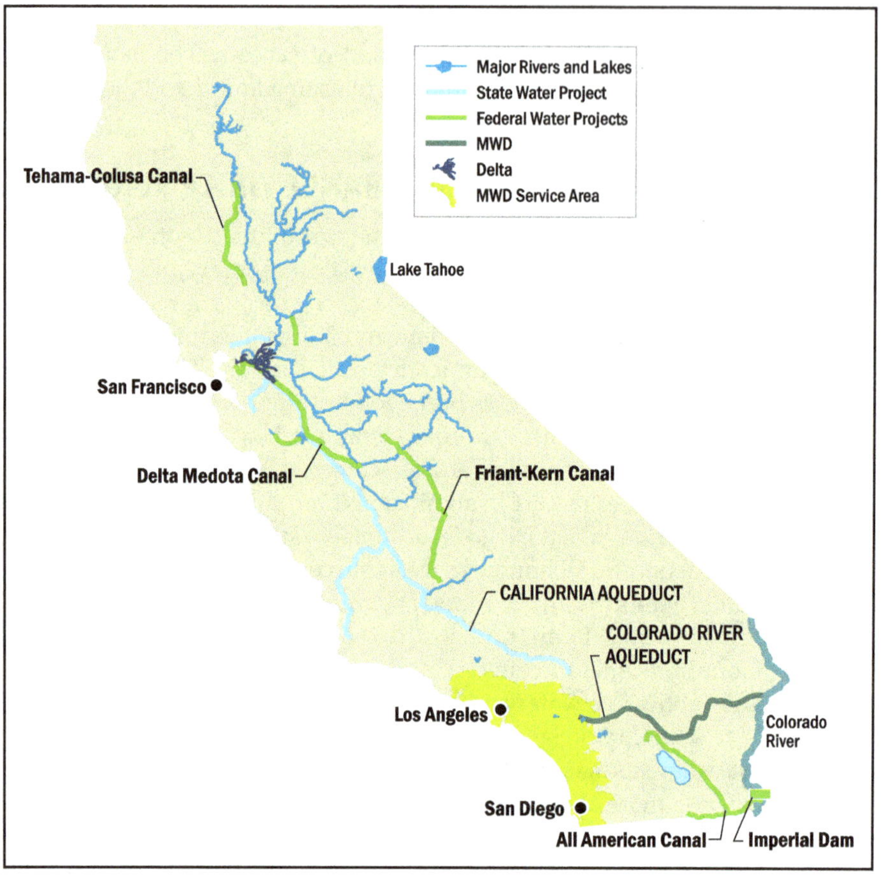

Source: *SDPUD 2012*

Figure 10-1 Sources of San Diego water

26 public water agencies and is the largest wholesale water agency in the nation. MWD was incorporated by the state legislature in 1928 to build the Colorado River Aqueduct, a facility it owns and operates. MWD also imports water from Northern California, originating at the Sacramento–San Joaquin Delta (Delta), through the State Water Project (SWP). MWD currently delivers approximately 1.8–2.1 million AFY (2.2–2.6 billion m^3/year) of imported water to its customers, but demands vary significantly with weather and economic conditions.

Local resources make up approximately 28 percent of the City's existing overall water supply, including savings from conservation. The SDPUD owns and operates an extensive raw water system that includes nine reservoirs, which captures local surface water for treatment at its three drinking water treatment plants (see Figure 10-1). The SDPUD has been investigating the feasibility of local groundwater and is pumping a small amount of groundwater that is conveyed to the City's water treatment plant for potable use. To offset potable system demands, the SDPUD has two water reclamation plants and a large-scale distribution system that delivers recycled water for nonpotable applications—primarily outdoor irrigation and also industrial cooling towers.

Water Resources Issues

With such heavy reliance on imported water, the City had to examine the various risk elements associated with that supply. A 12-year drought in the Colorado River Basin, more severe than any other measured in the 20th century, resulted in record lows in Colorado

River water levels. Water supply from the SWP has also been significantly reduced because of recent court restrictions to protect fisheries in the Delta and a prolonged statewide drought. The Delta, where pumping for both the SWP and federal water projects originates, is of particular concern because of the long-term degradation of habitat, risk of levee failures due to natural disasters, and long-standing conflicts between agricultural, urban, and environmental interests. These strains on MWD's water supply sources have resulted in imported water allocation limits to MWD's member agencies for the first time since 1991. While other SDCWA supplies help offset some of these shortages, San Diego is located at the "end" of imported water pipelines and is vulnerable to future restrictions—which are projected to be worse in the future with climate change.

Water quality for both the SWP and the Colorado River are also of significant concern to the City, because water quality tends to degrade over long distances of water conveyance. Salinity is of particular concern because source water that is high in salinity can cause damage to residential and industrial plumbing fixtures, destroy crops, and prevent the City from using its reclaimed water to its fullest potential.

In addition to reliability issues, the cost of imported water has increased significantly and is expected to continue to increase into the future. From 2007 to 2012, MWD's imported water costs increased more than 12 percent annually, and MWD projected its 2014 full-service water rate to be 7 percent greater than its 2012 rate. Going forward, SDCWA untreated water rates are expected to double in the next 20 years; however, projections are uncertain and it is possible that increases could occur even faster. Overall, projected imported water costs are expected to continue to increase faster than inflation, particularly given the major multibillion dollar improvements needed to solve the Delta conflicts. Finally, there are numerous institutional issues that cause uncertainty in imported water supplies, which mainly center on MWD's rate structure, wheeling, and drought allocation. The MWD and the SDCWA disagree on many of these issues.

In addition to imported water concerns, the City must consider the key initiatives and issues regarding development of local water resources with respect to

- limited local freshwater resources;
- meeting emergency storage requirements;
- water quality of local and imported water;
- maximizing reuse and minimizing wastewater discharges;
- meeting state conservation mandates; and
- addressing climate change effects on water demands, local water supplies, and availability of imported water.

Long-Range Water Resources Plan Process

The 2012 Long-Range Water Resources Plan (LRWRP) process was developed using an open, participatory planning process. Stakeholder collaboration was essential to the success of this plan's development. Throughout the process, the terminology found in Figure 10-2 was used.

At the start of the 2012 LRWRP process, the City formed a Stakeholder Committee that represented a wide range of interests and backgrounds to help guide the development of the plan. Members of the Stakeholder Committee included individuals from the following groups: San Diego County Taxpayers Association, Independent Rates Oversight Committee, San Diego Regional Chamber of Commerce, Building Industry Association of San Diego, San Diego Coastkeeper, American Society of Landscape Architects, City Representative to the SDCWA Board, and the San Diego Section of the American Planning

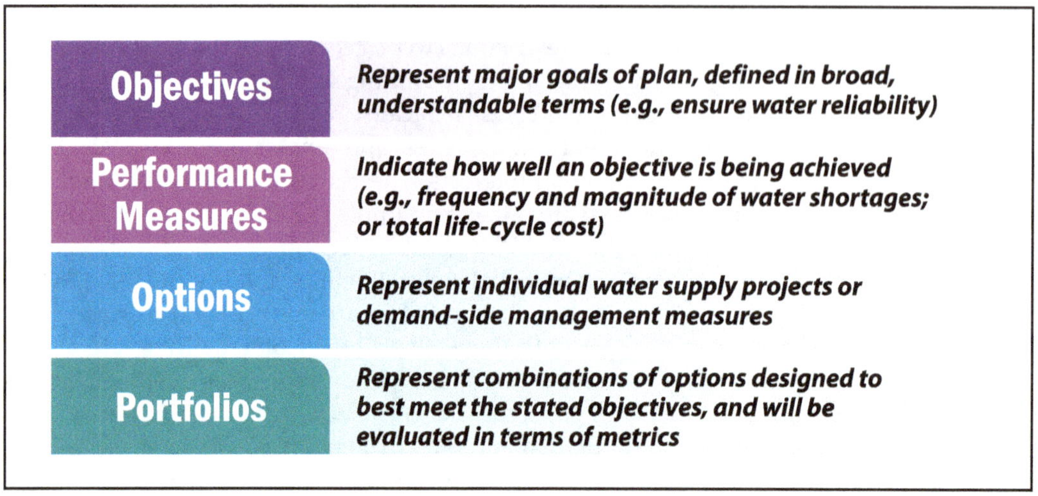

Source: *SDPUD 2012*

Figure 10-2 Terminology used with stakeholder involvement

Association. A total of five 2012 LRWRP Stakeholder Committee meetings were held over a period of approximately one year with the following goals:

1. Establish the need for 2012 LRWRP and define goals and objectives.
2. Review water supply and conservation options, and develop initial portfolios.
3. Review the evaluation of portfolios and provide comments on how to move forward.
4. Review the adaptive management approach and obtain consensus on recommended strategy.
5. Review the draft 2012 LRWRP and provide final comments/suggestions.

After the third stakeholder meeting, one-on-one conversations were held with all stakeholders in which several questions were asked about the process thus far, and to ascertain whether the information provided to date was understandable, fair, objective, and useful in the context of developing the 2012 LRWRP. The results of these one-on-one stakeholder conversations indicated that stakeholders felt the process was informative, objective, and very useful in the context of preparing the 2012 LRWRP. The stakeholders were also in general agreement with the overall evaluation process and findings to date.

2012 LRWRP Objectives

Planning objectives indicated that the goals of the 2012 LRWRP be stated in broad, understandable terms. A brainstorming session was held with the Stakeholder Committee early in the process to identify and define the objectives. After some refinement, the 2012 LRWRP objectives were defined:

- Provide reliability and robustness.
- Manage cost and affordability.
- Maximize efficiency of water use.
- Provide for scalability of implementation.
- Maintain current and future assets.
- Provide for local control/independence.
- Maximize project readiness.

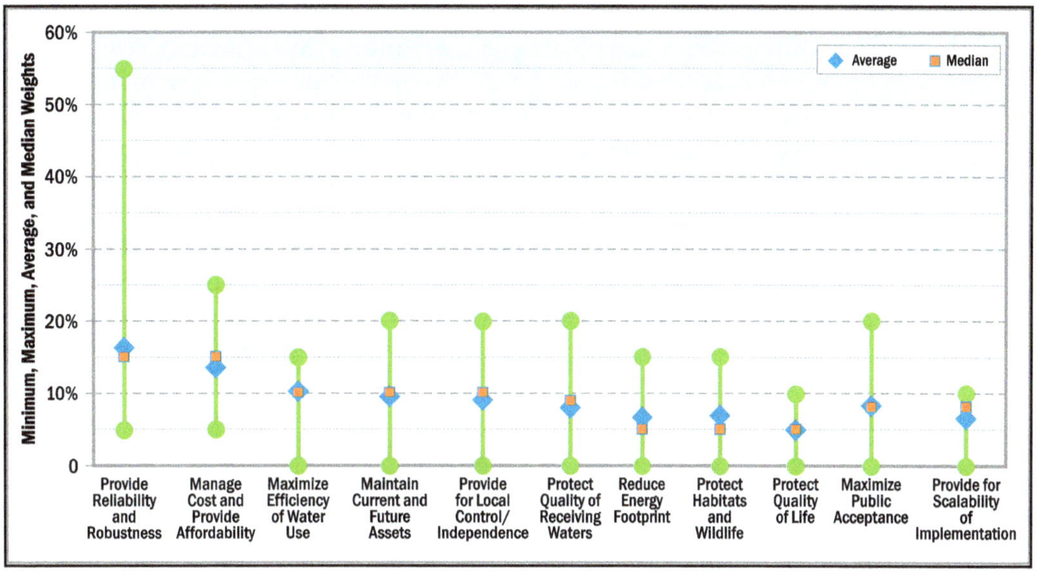

Source: *SDPUD 2012*

Figure 10-3 Comparison of objective weights for the 2012 LRWRP Stakeholder Committee

- Protect quality of life.
- Protect habitats and wildlife.
- Reduce energy footprint.
- Protect quality of receiving waters.

Because not all stakeholders viewed these objectives in the same way, with respect to relative importance, a weighting exercise was conducted. Each stakeholder was given a weighting form to allocate 100 points among the objectives in relation to their importance. The result of this weighting is shown in Figure 10-3, where the vertical lines represent the range of weights assigned to each objective by all stakeholders, the diamond marker indicates the average (or mean) weight, and the square marker shows the median (half of respondents fall above, half fall below) for all stakeholders. Except for the reliability objective, the overall spread of the minimum and maximum weights are relatively close, indicating general consensus among the group on the importance of the objectives.

For each objective, several sub-objectives and performance measures were established that would indicate how well the objective was being achieved. In total, more than 20 separate performance measures were established for the objectives and would be used to evaluate alternatives to develop the 2012 LRWRP strategy.

2012 LRWRP Evaluation Process

Because of the complexity of the City's water, recycled water, and wastewater systems, and because of the highly variable nature of local and imported water supplies, a sophisticated evaluation process was required to develop the 2012 LRWRP (see Figure 10-3). This evaluation process relied on engineering expertise, past technical studies conducted by the City on various water supply alternatives, sound water demand forecasting, and the use of simulation models and decision tools.

During development of the 2002 LRWRP, the City made a significant investment to develop a comprehensive system simulation model called San Diego Simulation

(SDSIM). This model simulates water demands, local supplies, and imported water availability under various hydrologic conditions and scenarios. The model routes local and imported water through the City's complex delivery system, matching water demands along the way. The model tracks life-cycle costs (i.e., capital and operations and maintenance [O&M]), water quality, storage operations, and other metrics such as greenhouse gas emissions. For the 2012 LRWRP, the model was enhanced to estimate the effects of climate change on system reliability.

Because SDSIM produces output in various raw measurements (e.g., cost in dollars, flows in acre-feet [or cubic meters] per year, greenhouse gas emissions in metric tons), a decision software tool called Criterion Decision Plus (or CDP) was needed to standardize the outputs, align them to the planning objectives, and apply stakeholder weights for the objectives. The process used by CDP to rank alternatives is called "multi-attribute rating," a standard approach used for multicriteria decision making. Because no single water supply option can meet all of the goals of the 2012 LRWRP, options were combined in different ways to form portfolios. (Much like a stock portfolio, a resource portfolio represents a diversified approach to meeting the plan's objective.) Stakeholders helped define how the portfolios would be assembled from the list of options. These portfolios were then input into SDSIM. Each portfolio was run against multiple hydrologic scenarios (wet, normal, dry, critically dry) and the output was synthesized and input into CDP. CDP standardized all of the raw performance metrics from SDSIM so they could be added together, aligned to objectives, and weighted. CDP produced an overall weighted score so portfolios could be ranked. Several initial portfolios were tested first, and then, based on their performance, several hybrid combinations were developed.

Water Supply Options

Using information from a variety of sources, more than 20 water supply and conservation options were identified for the 2012 LRWRP. Because of the high-level, strategic character of this plan, these options were considered conceptual in nature—although some have more technical robustness than others. Before any specific project recommendations are made concerning these options, further detailed study would be required including engineering design, environmental/permitting, refined cost estimation, facility layouts, and rate justification.

Table 10-1 presents the range of water supply and conservation options and their planning-level supply yields and costs in today's dollars. The unit cost, expressed as dollars per acre-foot of supply produced, incorporates the cost of supply development (capital, O&M, and energy); water distribution; and wastewater conveyance, treatment, and discharge. This reflects the full cost and puts all options on the same level.

In addition to costs, another important consideration for the City in developing the 2012 LRWRP was greenhouse gas emissions as required by the California Global Warming Solutions Act (AB 32, 2006). The City's reliance on imported water that originates hundreds of miles away, and requires energy-intensive pumping, contributes significantly to greenhouse gas emissions. The greenhouse gas emissions per acre-foot of water supply were calculated for each of the options.

Resource Portfolios

As stated previously, because no single supply option can meet all the objectives of the 2012 LRWRP, portfolios representing various combinations of options were developed. Stakeholders helped define how the initial portfolios were developed and which options were to be included. Based on the initial evaluation of these portfolios, the City along with its consultant team developed several hybrid portfolios, summarized as follows:

Table 10-1 Range of options considered

Supply Category	Number of Options	Range of Supply Yields (AFY)	Range of Unit Cost ($/AF)
Conservation Increase local conservation programs within San Diego	2	6,750–14,150	$200–$500
Groundwater Increase groundwater supply within San Diego	6	500–10,000	$1,400–$4,100
Recycled Water for Non-Potable Reuse* Increase reuse of treated wastewater for nonpotable applications	2	2,700–5,500	$2,100–$10,900
Recycled Water for Indirect Potable Reuse* Reuse of purified treated wastewater for indirect potable reuse	3	16,800–89,600	$2,100–$4,700
Rainwater Harvesting Capture of urban runoff for water supply	2	100–416	$6,400–$19,800
Graywater Non-sewage, onsite household wastewater that can be reused for nonpotable uses	1	2,575	$13,500
Ocean Desalination Pay higher purchase cost to SDCWA in exchange for more reliable ocean desalination water	1	10,000	$3,100
Imported Water Increase imported water purchases from SDCWA	1	As Needed and Available	$1,700
Other Concepts Considered Other groundwater, recycled, imported, etc.	6	NA	NA
Total:	24	100–56,000	$200–$19,800

AF = acre-feet AFY = acre-feet per year NA = not available

*Unit costs represent those used at the time of 2012 LRWRP analyses and are not based on most recent information available.

Source: *SDPUD 2012*

- *Baseline (Status Quo).* Only existing water supply and conservation is included in this portfolio. Over time, the reliance on imported water from the SDCWA will increase to meet growing water demands.
- *Maximize Reliability.* Options included in this portfolio are those that have little to no hydrologic variability (and therefore are not subject to droughts or climate change) and are owned/operated by the SDPUD or the SDCWA. Options that rely solely on consumer behavior or customer maintenance are not included because they are not as reliable into the future.
- *Minimize Cost.* Options included in this portfolio are those that have a unit cost (dollars per acre-foot) less than the projected cost of imported water from the SDCWA.
- *Minimize Local Environmental Impacts.* Options included in this portfolio are those that produce lower amounts of greenhouse gases (compared to imported water), those that have minimal or easily mitigated habitat impacts, and those that improve receiving water quality (rivers, streams, bays, and natural groundwater).
- *Maximize Local Control.* Options included in this portfolio are those in which SDPUD and the City have control over in terms of cost, development, and operations into the future.
- *Maximize Water Use Efficiency.* Options included in this portfolio are those that increase the efficiency of how water is used in the service area, including all levels of conservation, reuse, and rainwater harvesting.

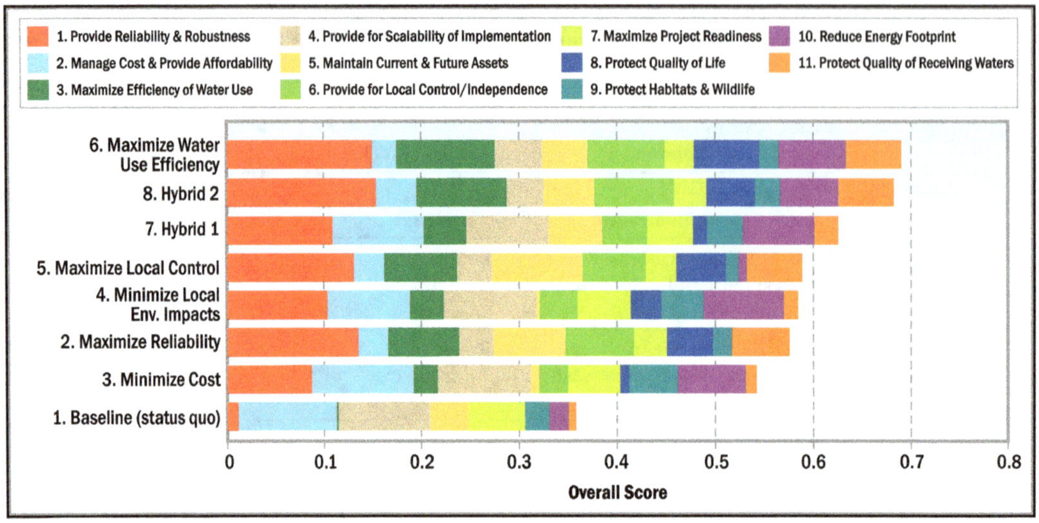

Source: *SDPUD 2012*

Figure 10-4 Portfolio rankings with average Stakeholder Committee weights

- *Hybrid 1.* This portfolio builds off the Minimize Cost portfolio by adding the Phase 1 Indirect Potable Reuse project.
- *Hybrid 2.* This portfolio builds off the Maximize Water Use Efficiency portfolio by adding groundwater projects, but removing nonpotable reuse with satellite treatment plants, graywater, and centralized stormwater capture.

Evaluation and Ranking of Portfolios

Portfolio ranking. More than 20 performance measures were used to comprehensively evaluate each of the portfolios. These performance measures were aligned to planning objectives. Raw performance scores for reliability, cost, water quality, greenhouse gas emissions, and other metrics were standardized to a unitless scale so they could be added together. Then portfolios were ranked based on their performance in meeting the objectives and the relative importance (or objective weight). Figure 10-4 presents the ranking of the portfolios based on the average stakeholder objective weights. The longer the color bar segment, the better the performance in achieving each objective.

Sensitivity. Sensitivity analyses were performed to assess whether the rankings are robust under different objective weighting scenarios. In addition, because long-term planning requires forecasting variables that are uncertain, sensitivity analyses were performed to assess the rankings under future uncertain conditions. The following sensitivity analyses were performed:

1. **Revised objective weights (from Figure 10-4).** Reliability and cost objective weights were revised from 17 percent and 14 percent, respectively, to 23 percent weight each, and all other objectives were given a 6 percent weight.
2. **Delta fix.** No imported water shortages would be projected, although this assumes higher imported costs.
3. **Higher energy costs.** Assume 30 percent higher energy costs, which affect cost of operations.
4. **Lower treatment technology costs.** Assume 30 percent lower operation cost for advanced treatment technologies used for indirect potable reuse and brackish groundwater treatment.

	6. Max Water Use Efficiency	8. Hybrid 2	7. Hybrid 1	5. Max Local Control	4. Min Local Env Impacts	2. Max Reliability	3. Min Cost	1. Baseline (status quo)
Baseline	1	2	3	4	5	6	7	8
Objective Weights	2	1	3	5	7	4	6	8
Delta Fix	1	2	3	5	4	6	7	8
Energy Cost	1	2	3	4	5	6	7	8
Treatment Tech	1	2	3	4	5	6	7	8
Objective Weights + Delta Fix	3	1	2	6	4	7	5	8

Source: *SDPUD 2012*

Figure 10-5 Portfolio rankings under sensitivity conditions

5. **Revised objective weights and Delta fix.** This combines the first two sensitivity conditions.

The findings for each weighting sensitivity scenario are summarized in Figure 10-5. The columns of the table represent the portfolios, the rows represent the sensitivity scenarios, and the number shows the rank order of the portfolio (1 being the best). The weighting scenarios are compared to the baseline rankings. The sensitivity results show that the top three portfolios are consistently Maximize Water Use Efficiency, Hybrid 2, and Hybrid 1 under all the sensitivity scenarios.

Climate Change Adaptation

In developing any long-term water supply plan, water utilities in California should consider the potential effects of climate change. Climate change is important to long-term planning because it could change hydrologic conditions and increase the need for new programs or investments that are resilient to climate variability in the future. While the potential effects of climate change are a global-scale concern, they are particularly important in the Pacific Coast region of the United States, which is one of the areas showing the most change with the greatest implications to water resources.

For San Diego, climate change can affect water demands, local water supplies, and availability of imported water. To analyze climate change effects, two climate change scenarios were used that are based on general circulation models downscaled to the San Diego region by the Scripps Institution of Oceanography. These two climate scenarios represent a realistic lower and upper range of potential effects. Figure 10-6 summarizes the effects of these two climate change scenarios on local weather, local surface water, and imported water availability.

The effects that changes in local weather will have on the City's water demand ranges from a 0.5 to 4 percent increase by 2035. This coupled with changes in local surface water and imported water availability results in changes in water supply reliability for the City. Under no climate change, overall water shortages for the City are estimated to occur roughly one in three years (33 percent) and the maximum shortage would be approximately 70,000 AFY (86.3 MCM/year), which is approximately 23 percent of projected 2035 baseline demands. The Geophysical Fluids Dynamics Laboratory climate scenario, characterized as being very warm and dry, would result in some type of shortage happening every year with a maximum shortage of approximately 100,000 AFY (123.3 MCM/year). The National Center for Atmospheric Research climate scenario, characterized as being

Impact by 2035	Climate Scenario 1 (GFLD CM2)	Climate Scenario 2 (NCAR PCM)
Local Temperature (Change from historical average)	+5% ↑	+3% ↑
Local Rainfall (change from historical average)	+1% ↔	+13% ↑
Local Water Demands (increase from historical normal)	+3.8% ↑	+0.5% ↔
Local Surface Water (change from historical average)	-7% ↓	+20% ↑
Imported Water (change from historical *normal* year)	-14% ↓	-8% ↓
Imported Water (change from historical *wet* year)	-6% ↓	-3% ↓

Bad Outcome Neutral Outcome Good Outcome

GFDL = Geophysical Fluids Dynamics Laboratory; NCAR = National Center for Atmospheric Research

Source: *SDPUD 2012*

Figure 10-6 Potential 2035 climate change effects for San Diego water supplies

warm and wet, would result in shortages occurring every two in three years (66 percent) with a maximum shortage of approximately 80,000 AFY (98.7 MCM/year).

These climate scenarios were also tested against the Hybrid 1 and Hybrid 2 portfolios to evaluate how well they provided adaptation to climate change. Because Hybrid 2 included more phases of indirect potable reuse (which is very resilient to climate change), it provides almost 100 percent adaptation to climate change effects. Hybrid 1 provides solid adaptation benefits, but not as great as Hybrid 2, thus greatly reducing the frequency and magnitude of water shortages.

Ranking Conclusions and Stakeholder Recommendations

Based on the conclusions from the portfolio evaluation and climate change analysis, the Stakeholder Committee recommended that the City develop a long-term strategy around the three top-ranking portfolios using an adaptive management framework.

Adaptive Management and 2012 LRWRP Strategy

As discussed previously, eight portfolios were evaluated against more than 20 performance metrics. The portfolios were ranked in terms of their cumulative performance. The portfolios were tested against five sensitivities and then re-ranked. Based on these rankings and their climate change adaptation benefits, three portfolios consistently ranked highest—Hybrid 1, Hybrid 2, and Maximum Water Use Efficiency (maximum efficiency). Figure 10-5 summarizes the resource options that are included in these three top-scoring portfolios.

Because there are more resource options in common among the three top-ranking portfolios, the City developed a strategy that uses adaptive management for implementing options.

Adaptive management. For purposes of the 2012 LRWRP, *adaptive management* is defined as a process in which options are implemented in a phased and incremental manner based on the outcome of identified future conditions or "risk triggers." Adaptive management balances the cost of option implementation with the risks of no action with four major steps:

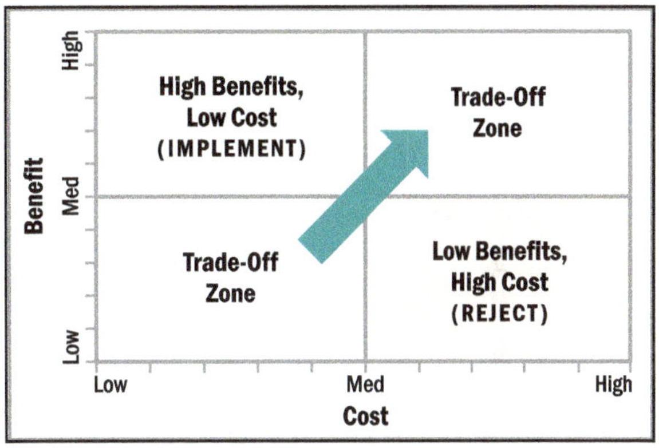

Source: *SDPUD 2012*

Figure 10-7 Benefit–cost trade-off analysis

1. Analyze trade-offs in terms of benefits and costs for the top-scoring portfolios.
2. Determine no-regret options that perform the best under most scenarios or sensitivities.
3. Establish near-term actions required to implement the no-regret options.
4. Develop risk triggers (points of uncertainty, along with projected possible outcomes) to determine future alternative paths of implementation for the remaining long-term options.

Although the three top-scoring portfolios have more in common than not in common, the total life-cycle costs in present value terms are very different. As such, it is important to analyze the trade-offs in terms of benefits and costs. Three main benefits were aggregated from key objectives and performance measures; these three benefits were then analyzed in comparison with costs:

1. Supply reliability benefit—representing drought supply protection, emergency storage, and resiliency to climate change
2. Environmental benefit—representing local habitat effects and greenhouse gas emissions
3. Water quality benefit—representing receiving water quality (from wastewater and stormwater discharges), groundwater basin quality, and salinity in drinking water

For ease of comparison, all three benefits were normalized to a score ranging from 0.0 (no benefit) to 1.0 (maximum achievable benefit). For each of the benefit categories, the three top-scoring portfolios and the status quo (do-nothing alternative) were plotted on a chart with *benefit* on the vertical axis and *total life-cycle cost* on the horizontal axis. This produces a four-quadrant analysis that indicates the overall benefit–cost trade-off as shown in Figure 10-7.

Portfolios that fall into the upper-left quadrant (high benefits, low cost) are the best alternatives and should be implemented with priority. Alternatively, portfolios that fall in the lower-right quadrant (low benefits, high cost) should be rejected outright. But it is not so apparent whether to implement portfolios that fall in the lower-left (low benefits, low cost) or upper-right (high benefits, high cost) quadrants because this is where trade-offs are made. For example, is it worth increasing supply reliability by 30 percent for a 40 percent increase in cost?

Figure 10-8 presents the results of the trade-off analysis for supply reliability benefit. What is shown in this figure is that moving from the status quo to Hybrid 1 greatly

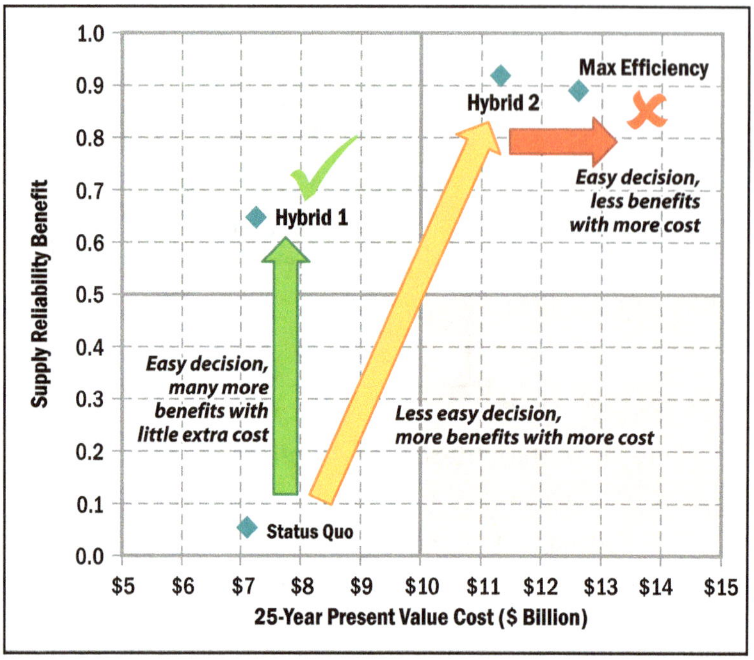

Source: *SDPUD 2012*

Figure 10-8 Reliability trade-off analysis

improves supply reliability for very little additional life-cycle cost. In fact, Hybrid 1 has a 10-fold increase in reliability benefits for only a 3 percent increase in cost compared to the status quo. From a supply reliability perspective only, implementing Hybrid 1 over the status quo is an easy decision, but moving from the status quo to Hybrid 2 is not as easy a decision. Supply reliability is increased 14-fold, but cost increases by almost 60 percent. Furthermore, the decision to not implement the Maximum Efficiency portfolio is an easy decision because it does not produce more supply benefit than Hybrid 2 but costs more.

Based on the other trade-off analyses, no-regret options were identified. These no-regret options show benefits under a wide range of uncertainty and therefore should be prioritized in the near term. These no-regret options include the following:

- Additional water conservation
- Additional groundwater investments
- Phase 1 indirect potable reuse
- Rainwater harvesting

Beyond the near term, other options should be implemented based on outcomes of future triggers. The Stakeholder Committee identified the triggers shown in Figure 10-9 for consideration in the 2012 LRWRP strategy.

Based on these risk triggers, a near-term and long-term adaptive management strategy for the 2012 LRWRP were developed. Figure 10-10 presents the near-term strategy (2012–2020), while Figure 10-11 presents the long-term adaptive element of the strategy (2020–2035).

This adaptive management strategy shows that there are many cumulative outcomes (following the path of the adaptive decision tree) in which no additional investments beyond those made in the near term (no-regret actions) are needed. In other words, Hybrid 1 is the only portfolio needed to be implemented. However, there are also several cumulative outcomes in which Indirect Potable Reuse (IPR) Phase 2/3 and additional groundwater are required (Hybrid 2).

Major Risk Trigger	Uncertainty	Implication or Impact
1. Acceptance	Permitting	Key to implementation of indirect potable reuse
	Customer Acceptance	Key to implementation of indirect potable reuse
	City Council Approval	Needed for large capital expenditures including funding
2. Cost	MWD/SDCWA Water Rates	If imported water rates are higher than expected, additional phases of IPR become more favorable
	Grant Funding	Grants could lower capital costs of more expensive groundwater and indirect potable reuse projects, making them more favorable
	Technology Improvements	Advancements in membrane technology could reduce costs of indirect potable reuse and brackish groundwater desalination
3. Imported Supply	Delta Fix	A Delta fix would improve imported water reliability, making investments beyond Hybrid 1 less favorable
4. Climate Change	Impact to Demands and Supplies	Would increase water demands and reduce water supplies making Hybrid 2 investments more needed
5. Direct Potable Reuse	Regulatory Approval	If DPR is approved by California regulatory agencies, and publicly accepted as well, the City may wish to consider DPR instead of Phase 2 and 3 of IPR

Source: *SDPUD 2012*

Figure 10-9 Risk triggers and implications

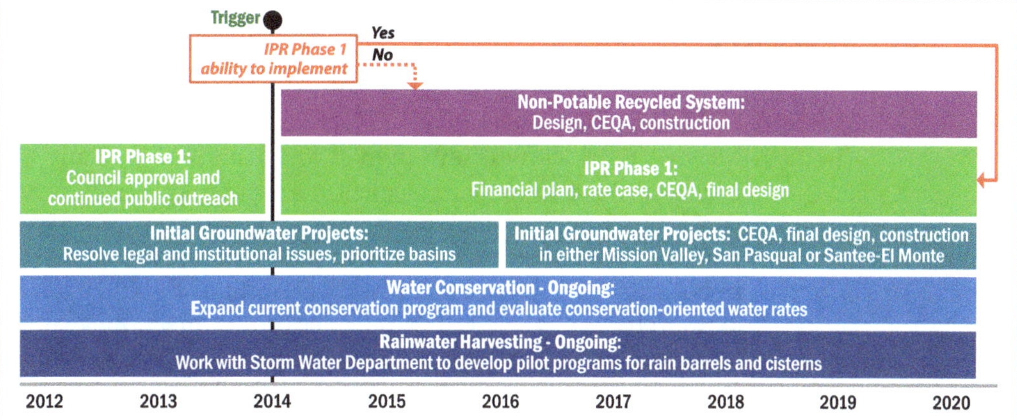

CEQA = California Environmental Quality Act

Source: *SDPUD 2012*

Figure 10-10 Near-term resource strategy

The City should be monitoring the risk triggers as well as water demand forecasts in five-year cycles, in concurrence with the City's Urban Water Management Plan. The risk triggers can be updated or changed depending on future circumstances. In addition, actions that are considered long-term strategies may be needed sooner rather than later. For example, if it is determined that climate change consequences to supply reliability are occurring sooner than expected, the timing of long-term actions should be moved up.

This adaptive approach will give the City a flexible road map for implementing its 2012 LRWRP, which emphasizes cost effectiveness, benefits to supply, water quality and environment, and minimizes risk.

In the long term, the outcome of additional triggers—which include Delta "fix," direct potable reuse regulations, climate change effects, or others—would indicate whether the City should move forward with implementing the additional elements included in the Hybrid 2 portfolio. However, if many of these triggers do not unfold as predicted, then the City would either stay the course with near-term actions (Hybrid 1 portfolio) or update the LRWRP as appropriate. The near-term and long-term options will reduce the City's reliance on imported water and potential future water supply shortages.

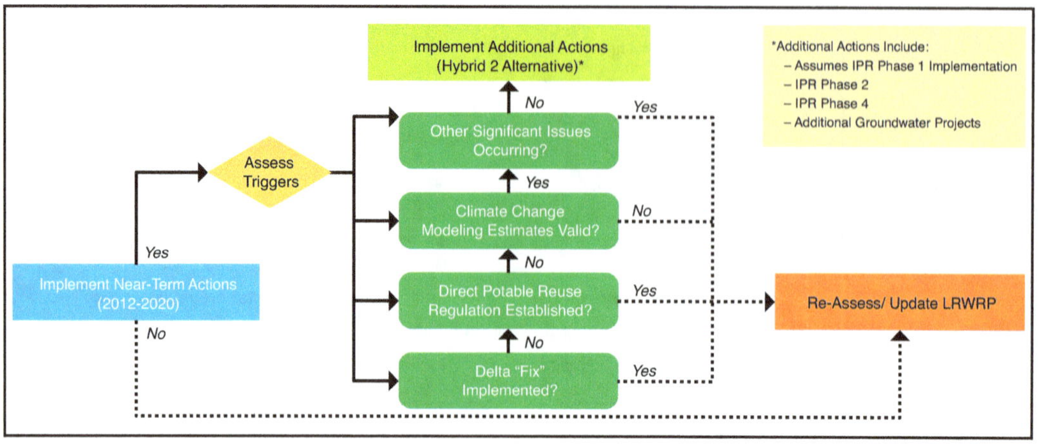

Source: *SDPUD 2012*

Figure 10-11 Long-term resource strategy

Recommendations

The following recommendations are made for the 2012 LRWRP and its implementation:

1. Move forward with the near-term actions identified in Figure 10-10. The near-term actions include steps to implement no-regrets options, which include additional water conservation, initial groundwater projects, rainwater harvesting, and Phase 1 of IPR. The next five years will be important in obtaining required permitting and public support for Phase 1 of the IPR project and resolving institutional, legal, and cleanup issues regarding groundwater development. By Year 2014, trigger should be assessed, which will indicate whether the City should proceed with IPR Phase 1. If the IPR project cannot be permitted and publically supported, or cannot receive City Council approval, then the City should assess whether to move forward with a more aggressive expansion of its nonpotable recycled water system.
2. Re-assess all 2012 LRWRP triggers and water demand forecast in 2015, in concurrence with the City's 2015 Urban Water Management Plan.
3. Move forward with an adaptive management approach as depicted in Figure 10-9, using the risk triggers identified in this plan or revised triggers.
4. All triggers and outcomes should be re-assessed in concurrence with the City's Urban Water Management Plan schedule (2020, 2025, 2030, 2035).
5. Update the entire LRWRP in 10-year cycles, starting in 2020.

REFERENCES

AB 32. 2006. California Assembly Bill 32: Global Warming Solutions Act.

Bloetscher, F. 2011. *Utility Management for Water and Wastewater Operator.* Denver, Colo.: AWWA.

Dzurik, A.A. 2003. *Water Resources Planning.* Lanham, Md.: Rowman and Littlefield.

SDPUD (San Diego Public Utilities Department). 2012. *City of San Diego Public Utilities Department 2012 Long-Range Water Resources Plan.* San Diego, Calif.: SDPUD.

Appendix A

Preliminary Cost Guide for Water Supply Dams

For reconnaissance-level evaluations, the following guidance provides relative cost estimates when comparing dam alternatives. Guidance is presented for selecting earth fill and concrete gravity dams, including developing opinions of cost to provide a reasonable framework for initial site screening assessments. For simplicity, this approach is limited in scope; it does not replace the counsel and input of an experienced dam design professional.

Most new concrete gravity dams are constructed using roller-compacted concrete (RCC), because RCC is significantly less expensive than conventionally placed concrete for most applications. Therefore, the cost guidelines for gravity dams presented later in this appendix were developed assuming the use of RCC.

SELECTING THE APPROPRIATE DAM TYPE

A far more extensive set of site conditions, opportunities, and constraints than those presented in this appendix could be developed. However, the issues and considerations presented are generally sufficient to initially identify and assess a dam type that is conducive to most sites. Table A-1 presents key site characteristics and their importance in selecting an appropriate dam type.

If a key factor for one dam type is identified to have an overriding detrimental impact on technical viability or cost (e.g., bedrock at substantial depth precluding an RCC dam), the other dam type should be selected by default. It is also not uncommon to encounter dam sites that lack clear merits for either dam type. In such cases, the utility needs to assess both types.

Table A-1 Considerations for selecting dam type

Assessment Factor	Earth Fill Dam	RCC Gravity Dam
Foundation Conditions		
Overburden	Significant concerns if highly permeable to depth or if excessive settlement likely; overburden cutoff trench typically used.	Typically built on rock foundation, with overburden excavated to competent rock.
Bedrock quality	Of concern if shallow overburden with open bedrock fractures. Generally not a significant concern unless soil cover is either minimal or relatively pervious.	Rock surface needs to be relatively uniform (no overhangs or ledges) and surface fractures treated. Rock grouting is typically used to treat dam foundation.
Construction Materials and Access		
Borrow availability	Need large quantities of suitable earth borrow relatively close to the proposed fill placement site.	Either a local quarry site (rock crushing) or minable sands and gravels are needed for larger projects. Commercial aggregates common for RCC volumes smaller than 30,000 yd^3 (22,937 m^3).
Access, stockpiling, and staging areas	Typically require access for hauling, spreading, compaction equipment, and concrete transit mix trucks, plus nominal staging areas.	Access as for earth dam, plus staging for RCC mix plant, aggregate stock-piles, drilling and grouting equipment, conveyors, rock crushers, etc.*
Spillway Requirements		
Spillway design flood magnitude	Normally need to locate major spillways away from the dam. Requirement for large spillway can significantly affect layout, design treatments, and project costs.	Spillway facilities can normally be accommodated as part of the impounding structure and normally do not significantly affect project costs.
Site topographic setting	If necessary, a large remote spillway can be accommodated more readily in valleys with a mildly sloping abutment or a reservoir perimeter saddle area.	Because the spillway is readily incorporated into the dam, an RCC dam benefits from a deeply incised valley but can be used for a variety of site settings.

*A much smaller footprint is needed for an RCC dam.

The preliminary cost-estimating techniques used do not consider pumping stations or pipelines, water supply intake and outlet works, property acquisitions, reservoir clearing, structure relocations, and bridges, or any other considerations not directly related to the impounding structure. The utility will need to consider the magnitude of these and other costs separately.

ESTABLISHING DAM PROFILE LAYOUT

The topography along the dam axis (from US Geological Survey [USGS] 7.5 minute quadrangles or other available mapping) should be used to plot a profile of the valley as much as 20 ft (6.1 m) above the proposed reservoir pool elevation, with the top of dam set 10 ft (3 m) above the proposed operating pool elevation, as shown in Figure A-1.

PROFILE LAYOUT AND BASIS FOR ESTIMATING QUANTITY

An approximate quantity take-off is developed from the profile layout presented in Figure A-1. The utility should estimate the abutment and foundation excavation lines required to found the dam on firm bedrock for an RCC dam and on competent overburden for an earth dam (a minimum of 5 ft [1.5 m]). The base length and top length of the dam (L_b and L_t) and the abutment contact slopes should be set to best fit the dam profile geometry, as shown in Figure A-2.

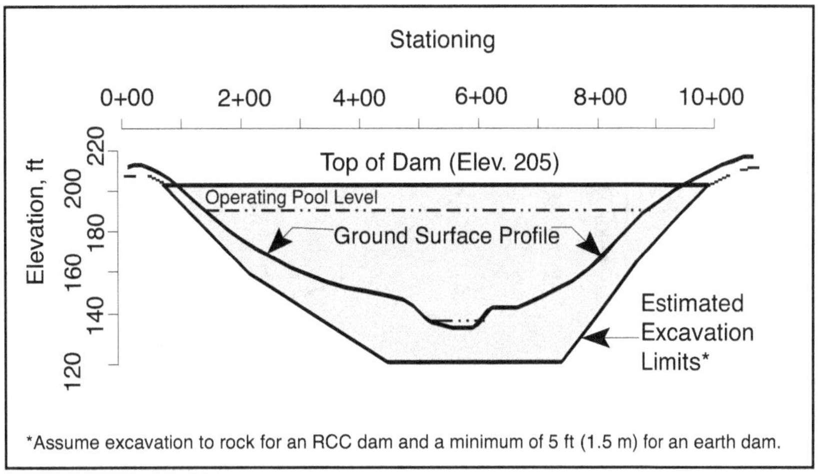

Figure A-1 Dam profile layout

County soil surveys will generally include the approximate thickness of soil strata and depths to bedrock for use in this assessment. Additionally, USGS geologic maps and bedrock outcropping noted during field reconnaissance can provide added guidance in this regard.

RCC Gravity Dam Assessment—Preliminary Layout and Opinion of Cost

For concrete gravity impounding structures, single-stage overflow spillways are typically incorporated into the dam. The cost of a spillway for a gravity dam generally is not an overriding factor affecting feasibility, because a large spillway can readily be incorporated into a gravity dam. In general, by applying only the larger cross-sectional area of the nonoverflow portions of the dam (see Figure A-3), the added volume of RCC will reasonably counterbalance the cost of treatments needed for safe flood passage over the spillway (a nominal spillway cost factor is added).

Developing RCC Dam Layout and Quantity

Once an outline profile of the dam is established, its volume can be estimated based on the characteristic dimensions illustrated in Figure A-3. Volume estimating procedures assume that the width of the top of the dam is 20 percent of the maximum dam height, the top of the dam elevation is 10 ft (3 m) above the operating pool, the slope of the downstream face of the dam is 0.8(h):1.0(v), and the slope intersects the top of the dam elevation at the upstream face (see Figure A-1). Should the top of the dam need to support a roadway, the additional costs for a wider top of the dam and a spillway bridge would need to be separately considered.

The required volume of RCC is approximated using the following equation:

$$V_{\text{RCC}} = 0.008 \times H^2 \times (2L_b L_t) \tag{A-1}$$

Where:

V_{RCC} = total estimated volume of RCC, in cubic yards

H = maximum height of dam, in feet (see Figure A-3)

L_b = total length of the base of the dam, in feet (see Figure A-3)

L_t = total length of the top of the dam, in feet (see Figure A-3)

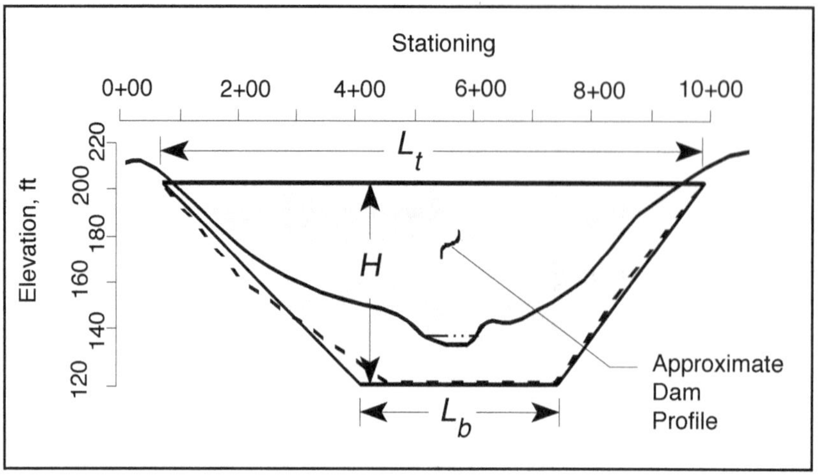

Figure A-2 Profile layout for estimating quantity

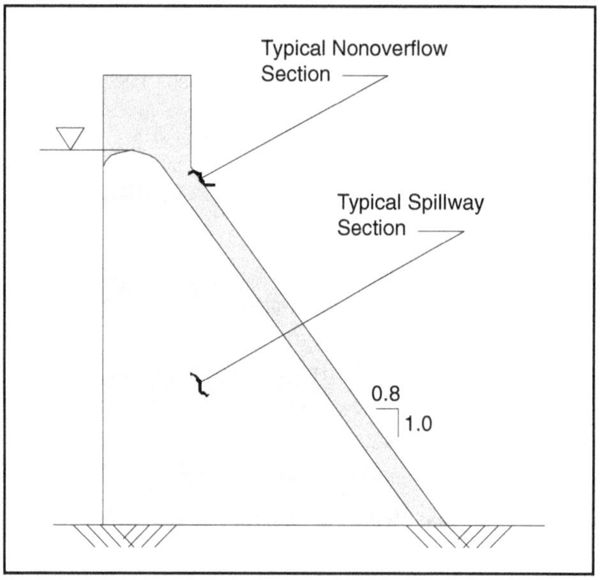

Figure A-3 Typical RCC dam geometry

RCC Dam—Opinion of Cost

Figure A-4 provides a graphical presentation of estimated RCC unit cost versus placement volume. High, low, and average cost curves are provided to reflect factors such as construction costs for the region in question, RCC aggregate availability, and access and staging areas for materials and equipment. Aggregate availability and processing costs can significantly influence the selected unit price for RCC. Additional cost factors are included for spillways and general mobilization. The estimated cost for an RCC gravity dam is computed as follows:

$$\text{cost (\$)} = V_{\text{RCC}} \times UC_{\text{RCC}} \times (1 + CF_s + CF_{gm}) \qquad \text{(A-2)}$$

Where:

V_{RCC} = RCC volume

UC_{RCC} = RCC unit cost

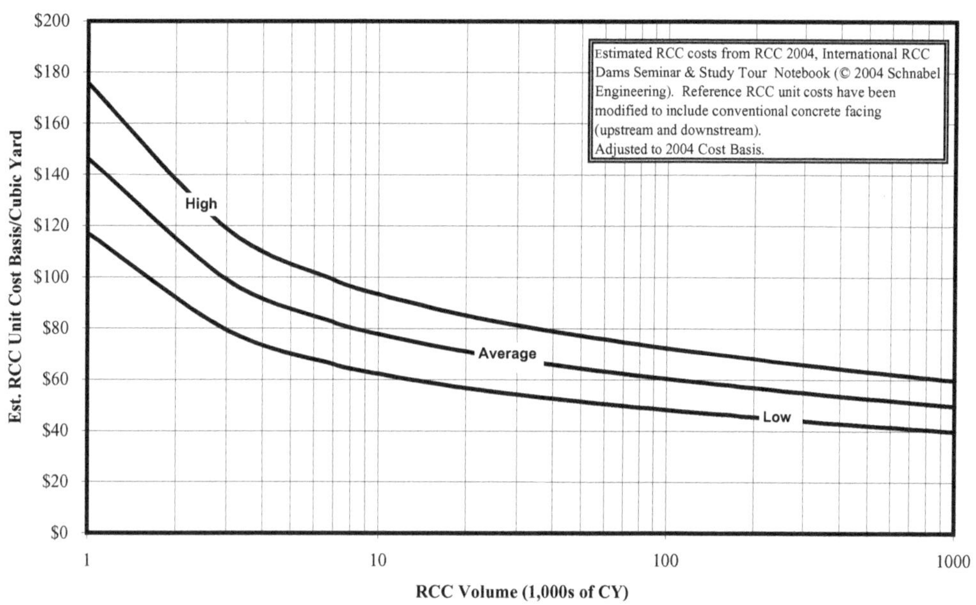

NOTE: Estimated RCC costs courtesy of the Portland Cement Association. Base costs have been modified to include estimated dam facing costs.

Figure A-4 RCC unit cost

CF_s = spillway size cost factor

CF_{gm} = cost factor for general mobilization that includes mobilization, site work, control of water, foundation treatment, and other incidental items

Utilities should apply considerable judgment when selecting cost factors.

CF_s should be selected in the range from 0.2 to 0.5, depending on the magnitude of the spillway design flood. In general, the smaller value would apply to nonmountainous drainage areas of 10 mi^2 (25.6 km^2) or less, with the high-end value for mountainous drainage areas of about 40 mi^2 (102.4 km^2).

For dams with moderate foundation excavation and treatment, or diversion and dewatering considerations, use a CF_{gm} of 0.4. Where a project will require extensive foundation excavation and treatment, and significant construction diversion and dewatering facilities, a CF_{gm} of 1.0 is recommended. Values should be selected to reflect likely project requirements.

This estimating technique does not include costs for stilling basins, pumping stations, pipelines, water supply intake and outlet works, property acquisition, reservoir clearing, structure relocations, and bridges, or any other facilities not directly related to the impounding structure. The magnitude of these and other costs will need to be considered on a site-specific basis.

EARTH FILL DAM ASSESSMENT—PRELIMINARY LAYOUT AND OPINION OF COST

Developing reliable cost opinions for earth fill dams is generally more complex and difficult than for RCC dams, because of the significant influence spillway facilities can have on project layout and cost. Four basic earth fill dam options are presented in Table A-2 to accommodate a range of site settings and spillway requirements.

Table A-2 Embankment dam alternative layout selection factors

Layouts for Earth Embankment Dams	SDF Normally Accommodated	Impact of Steep Abutments	Impact of Difficult Foundations	Local Borrow Material Availability	Range of Economical Dam Heights
Principal/ emergency spillway	Small to intermediate	Significant	Intermediate	Intermediate to significant	<100 ft (<30 m)
Central overflow spillway	Small to large	Nominal	Significant	Reduces need for borrow	<50 ft (<15 m)
Remote service spillway	Intermediate to large	Intermediate	Significant	Intermediate	>50 ft (>15 m)
Closed conduit spillway	Nominal	Nominal	Low	Negligible	Any

When developing concepts and layouts for an earth dam, the first consideration is the magnitude of spillway discharge required. Where a large spillway design flood (SDF) needs to be handled, spillway siting and layout often dictates the efficient use of site opportunities and approaches to treating other site conditions. If feasible, an RCC dam should be selected in these cases.

For most earth dams, constructing a major spillway within the dam footprint is also more costly and difficult than for an RCC dam. Where the SDF magnitude is more nominal (such as for a pumped-storage reservoir site on a small tributary watershed) or where the local topography and geology provide ample opportunity for locating an abutment or reservoir perimeter spillway, the layout of project facilities is more straightforward.

Selecting an Earth Dam Alternative

Four alternative dam concepts are presented for consideration. Based on the information presented in Table A-2, the embankment and spillway option considered most conducive to the site setting and SDF magnitude should be selected. Where two alternatives appear to be equally suited to the site conditions, the first listed alternative from Table A-2 should be selected.

Developing Earth Dam Layout and Quantity

Volume estimating procedures assume that the width of the top of the dam is 40 percent of the maximum dam height, the upstream and downstream slopes are 3 horizontal to 1 vertical, and a foundation cut-off trench is provided. The required volume of earth fill is approximated using the following equation:

$$V_{\text{earth}} = 0.05 \times H^2 \times (2L_b + L_t) \tag{A-3}$$

Where:

V_{earth} = total estimated volume of earth fill, in cubic yards

H = maximum height of dam, in feet (see Figure A-3)

L_b = total length of the base of the dam, in feet (see Figure A-3)

L_t = total length of the top of the dam, in feet (see Figure A-3)

Because the reduction in earth fill for a central overflow spillway is generally offset by the need for a more costly structural spillway, the utility should not reduce earth fill volumes for this option. Figure A-5 provides a graphical presentation of estimated earth

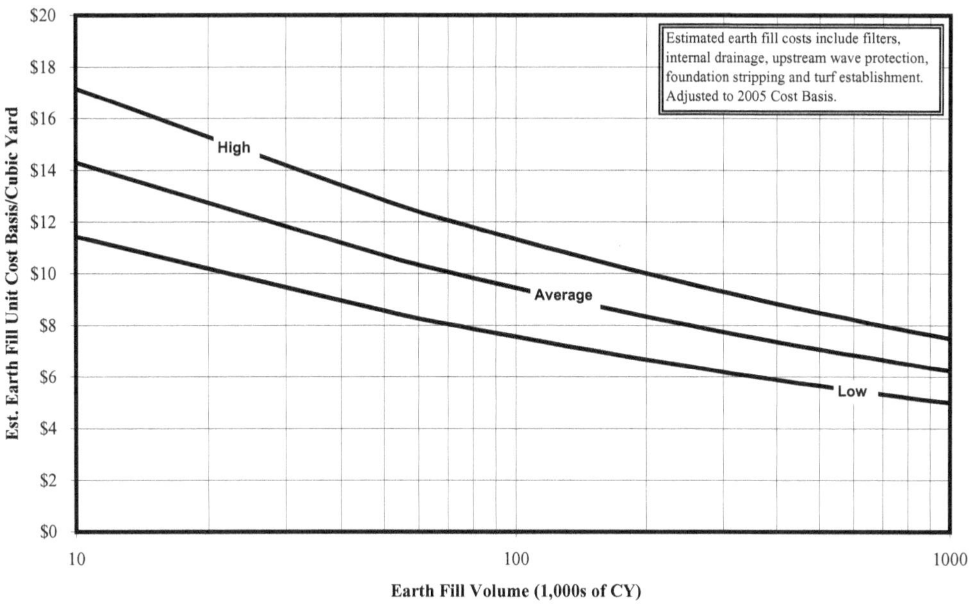

NOTE: Considerable judgment needs to be applied in selecting cost factors.

Figure A-5 Earth fill unit cost

fill unit cost versus placement volume. High, low, and average cost curves are provided to reflect factors such as construction costs for the region, earth borrow availability, and access and staging areas for materials and equipment. Additional cost factors are included for spillways and general mobilization.

EARTH DAM—OPINION OF COST

The estimated cost for an earth dam is computed as follows:

$$\text{cost (\$)} = V_{\text{earth}} \times UC_{\text{earth}} \times (1 + CF_s + CF_{gm}) \quad \text{(A-4)}$$

Where:

V_{earth} = earth fill volume

UC_{earth} = earth fill unit cost

CF_s = spillway size cost factor

CF_{gm} = cost factor for general mobilization that includes mobilization, site work, control of water, foundation treatment, and other incidental items

CF_s should be selected in the range from 0.5 (small) to 1.0 (intermediate) to 1.5 (large), depending on the magnitude of the SDF. In general, the smaller value would apply to nonmountainous drainage areas of 10 mi^2 (25.9 km^2) or less, with the high-end value for mountainous drainage areas of about 40 mi^2 (101.6 km^2). For pumped-storage projects on tributary streams less than 5 mi^2 (12.8 km^2), CF_s can be reduced to 0.3 (nominal).

For dams with moderate foundation excavation and treatment, and diversion and dewatering considerations, use a CF_{gm} of 0.4. Where a project requires extensive foundation excavation and treatment, and significant construction diversion and dewatering facilities, a CF_{gm} of 1.2 is recommended. Values should be selected to reflect likely project requirements.

This page intentionally blank.

Appendix B

A White Paper From the AWWA Source Water Protection (approved April 11, 1997)

STATEMENT OF PRINCIPLES

The American Water Works Association (AWWA) is dedicated to providing the public with an adequate supply of clean, safe drinking water. AWWA is committed to ensuring that water resources are managed in a manner consistent with the protection and enhancement of source waters for current and future supplies of drinking water. Source water protection (SWP) is a program of actions, policies, and practices to be undertaken by water suppliers, government agencies, institutions, and individuals to advance these goals.

AWWA promotes a multiple-barrier approach to providing safe drinking water that includes SWP, treatment as appropriate, distribution system maintenance, and monitoring. SWP may reduce health risks and treatment costs and improve finished water quality. SWP programs may also provide ancillary benefits of enhancing water quality for other users and improving the natural and aesthetic environments of communities. Accordingly, SWP should be pursued diligently for every water supply source.

SWP programs must be implemented in a context of supporting and competing public needs. They also must be flexible enough to address threats to source water quality and opportunities for improvement that vary from site to site and evolve over time. Regulatory programs and subsidies at all levels of government that are related to water resource protection should focus on current or potential sources of drinking water. In these programs, SWP goals should be added or elevated in importance.

Water suppliers, regulators, landowners, and municipalities share responsibility for protecting source water. Property owners must bear responsibility for preventing and abating pollution emanating from their holdings. AWWA supports the interests of water suppliers and of consumers whose health and welfare could be affected by unrestricted exercise of property rights upstream. However, AWWA recognizes the need to be sensitive to property rights and to avoid imposing undue burdens on parties who may be affected by SWP measures.

THE ROAD TO A SOURCE WATER PROTECTION PROGRAM

A basic premise for implementing SWP programs is the multiple-barrier approach to protecting water supplies and public health. By establishing multiple barriers that include SWP, treatment as appropriate, distribution system maintenance, and monitoring, water suppliers are able to ensure the quality and safety of drinking water for their consumers. SWP represents a first and most important step in safeguarding public water supplies.

Some common elements for successful SWP programs are: they account for local conditions, incorporate diverse interests, require commitment to the SWP process by all involved parties, and are sustainable over the long term. SWP requires a sustained commitment of policy, as well as financial and technical resources over a time span of decades, not just years. Some important water quality benefits of SWP may not be measurable in the short run. In addition, a long-term commitment is necessary to ensure the protection of high-quality water sources so they remain available for future generations.

One of the most difficult issues in an SWP plan is the establishment of equity in sharing the responsibility and expense of these programs. SWP efforts are often hampered by issues of who benefits and who pays. The following guidelines have been used to resolve these issues:

- Sources of pollution bear the responsibility for remediation; in other words, the polluter pays.
- Open and active communication, flexibility, and participation in the SWP process by involved parties can overcome actual and perceived imbalances of equity.
- Federal, state, and local resources can be applied to help address the equity issue.
- Consideration should be given to the value that SWP programs can provide to a community through environmental benefits—such as wildlife habitat and open space—as well as improved quality and quantity of available resources.
- Compensation for lost or diminished use of property because of SWP restrictions may be considered in some cases.

These guidelines can help balance the rights of property owners and others affected by SWP measures with the rights of consumers whose health and welfare depend on the quality of source waters that could be degraded by the exercise of unrestricted property rights.

DEVELOPING RESOURCES FOR SOURCE WATER PROTECTION

A challenge incumbent on successful, sustainable SWP programs is developing adequate structural and financial resources to support them. Some specific options include the following:

- State and federal governments tailoring legislative and regulatory agendas, resources, and programs to support SWP.
- State governments refocusing and allocating a portion of resources and funding to SWP. The states should secure adequate legislative and regulatory authority (e.g., planning and regulatory enforcement) for SWP programs. This could also include levies on polluters or pollutants (pesticides, herbicides, fertilizers, etc.), with the proceeds supporting cleanup efforts.
- Local governments supporting SWP with appropriate land-use management and regulatory enforcement and by encouraging support from local grassroots efforts, environmental groups, and community groups.
- Water suppliers taking an active role in protecting their source waters by providing organizational, technical, monitoring, and financial resources and by harnessing resources available from federal, state, and local programs and institutions, and from volunteers.
- Private organizations initiating SWP programs and participating in cost-sharing arrangements.

RECOMMENDATIONS

1. Water suppliers, regulators, and local landowners and municipalities share responsibility for SWP. Federal and state programs need to be tailored to support a local and regional approach to developing and implementing SWP programs and activities.
2. Recognizing that drinking water sources are becoming significantly polluted now, federal and state legislative and regulatory programs should be directed to stress the protection of water resources for current or potential drinking water supplies on a priority basis. SWP goals should be included in programs and, where already included, elevated in importance.

   Where necessary and appropriate, new or expanded regulatory SWP programs should be implemented for specific river basins, watersheds, or aquifers via state or regional initiatives. This implies an integrated look at all the activities within an aquifer or watershed to assess priorities and place priority on certain pollution protection programs that offer the best net economic and environmental benefits.
3. Water suppliers should develop written source water management plans to prevent or reverse water quality degradation. The SWP plans should delineate and characterize specific source water areas (watershed, wellhead, or recharge areas), identify threats to water quality, and provide a strategy for ongoing management of conditions and activities within these areas that may affect source water quality. The plans should also specify resource requirements for communications, implementation, and program assessment.

This page intentionally blank.

Appendix C

State Wellhead and Source Water Protection Contact List

For a current updated list of the State Wellhead and Source Water Protection Contacts, please go to www.epa.gov/sourcewaterprotection

For inquiries on national drinking water source protection programs, call the SDWA Hotline, 1-800-426-4791.

This page intentionally blank.

Index

NOTE: *f.* indicates figure; *t.* indicates table

L

M

N

O

P

R

S

T

U

V

W

Z

AWWA Manuals

M1, *Water Rates, Fees, and Charges,* #30001
M2, *Instrumentation and Control,* #30002
M3, *Safety Management for Water Utilities,* #30003
M4, *Water Fluoridation Principles and Practices,* #30004
M5, *Water Utility Management,* #30005
M6, *Water Meters—Selection, Installation, Testing, and Maintenance,* #30006
M7, *Problem Organisms in Water: Identification and Treatment,* #30007
M9, *Concrete Pressure Pipe,* #30009
M11, *Steel Pipe—A Guide for Design and Installation,* #30011
M12, *Simplified Procedures for Water Examination,* #30012
M14, *Backflow Prevention and Cross-Connection Control: Recommended Practices,* #30014
M17, *Installation, Field Testing, and Maintenance of Fire Hydrants,* #30017
M19, *Emergency Planning for Water Utilities,* #30019
M20, *Water Chlorination/Chloramination Practices and Principles,* #30020
M21, *Groundwater,* #30021
M22, *Sizing Water Service Lines and Meters,* #30022
M23, *PVC Pipe—Design and Installation,* #30023
M24, *Planning for the Distribution of Reclaimed Water,* #30024
M25, *Flexible-Membrane Covers and Linings for Potable-Water Reservoirs,* #30025
M27, *External Corrosion Control for Infrastructure Sustainability,* #30027
M28, *Rehabilitation of Water Mains,* #30028
M29, *Water Utility Capital Financing,* #30029
M30, *Precoat Filtration,* #30030
M31, *Distribution System Requirements for Fire Protection,* #30031
M32, *Computer Modeling of Water Distribution Systems,* #30032
M33, *Flowmeters in Water Supply,* #30033
M36, *Water Audits and Loss Control Programs,* #30036
M37, *Operational Control of Coagulation and Filtration Processes,* #30037
M38, *Electrodialysis and Electrodialysis Reversal,* #30038
M41, *Ductile-Iron Pipe and Fittings,* #30041
M42, *Steel Water-Storage Tanks,* #30042
M44, *Distribution Valves: Selection, Installation, Field Testing, and Maintenance,* #30044
M45, *Fiberglass Pipe Design,* #30045
M46, *Reverse Osmosis and Nanofiltration,* #30046
M47, *Capital Project Delivery,* #30047
M48, *Waterborne Pathogens,* #30048
M49, *Butterfly Valves: Torque, Head Loss, and Cavitation Analysis,* #30049
M50, *Water Resources Planning,* #30050
M51, *Air-Release, Air/Vacuum, and Combination Air Valves,* #30051
M52, *Water Conservation Programs—A Planning Manual,* #30052
M53, *Microfiltration and Ultrafiltration Membranes for Drinking Water,* #30053
M54, *Developing Rates for Small Systems,* #30054
M55, *PE Pipe—Design and Installation,* #30055
M56, *Nitrification Prevention and Control in Drinking Water,* #30056
M57, *Algae: Source to Treatment,* #30057
M58, *Internal Corrosion Control in Water Distribution Systems,* #30058
M60, *Drought Preparedness and Response,* #30060
M61, *Desalination of Seawater,* #30061
M63, *Aquifer Storage and Recovery,* #30063
M65, *On-Site Generation of Hypochlorite,* #30065
M66, *Actuators and Controls—Design and Installation,* #30066

This page intentionally blank.

www.ingramcontent.com/pod-product-compliance
Lightning Source LLC
LaVergne TN
LVHW080042210926
848481LV00011B/3262